"At last, a book written with the practitioner in mind. *Teaching for Understanding with Technology* will serve as an invaluable guide for educators everywhere. The authors speak in real terms, through the eyes of real students and teachers. The vignettes show how technology can empower and motivate both student and teacher. As a principal and instructional manager, I see this book as a must-have blueprint for all educators. I intend to purchase a copy for every staff member in my building."

Mary Skipper, headmaster, TechBoston Academy, Dorchester, Massachusetts

"If you have any doubts about the way in which technology can enrich the learning of all students, you are holding the book you need to read. Stone details the ways in which Kristi is reinventing learning in the age of technology and explains why this approach is so essential. What is remarkable is that when you treat first graders like graduate students they end up acting like them."

Margaret Riel, senior researcher, Center for Technology in Learning SRI, and visiting professor, Pepperdine University

"This book is about translation and transformation, using the new technologies to improve teaching and learning. It demonstrates how these new technologies, essential ingredients in education in the twenty-first century, can support teachers as they refine their practice, and make learning a deeper and more lasting experience students learn to understand. The book makes an elegant case for the appropriate and informed use of technology in our schools."

Isa Kaftal Zimmerman, director, Technology in Education Program, Lesley University

"This book is needed so that all educators will understand how to use the power of technology to propel teaching and student learning. Teachers need to understand how to create classroom projects with technology that build on the students' interests and extend those interests by having students communicating and collaborating with peers around the globe. This book can help teachers break through the barriers of integrating technology into their curriculum. Classrooms can then become learning environments where students reach out to their world and find their place in it."

Katherine Law, Seattle public school teacher and lead educational technologist

Teaching for Understanding with Technology

Teaching for Understanding with Technology

Martha Stone Wiske
with Kristi Rennebohm Franz and Lisa Breit

JOSSEY-BASS
A Wiley Imprint
www.josseybass.com

· W 5 6 9

The Jossey-Bass Education Series

Contents

Preface

The need for good public education has never been greater than in today's complex, interdependent, and rapidly changing world. At the same time, pressures on schoolteachers and administrators have rarely been as intense. Educators need ways to specify clear goals and to coordinate their efforts coherently to help all students succeed. New information and communication technologies can be valuable resources in developing effective strategies for promoting ambitious learning in public education. This book presents a detailed language and a practical structure—the Teaching for Understanding framework—for guiding educators through the maelstrom of competing pressures toward effective teaching and learning with new technologies.

Background of the Problem

Policymakers and politicians, business leaders and parents, educational researchers and developers—all have their own particular and often competing expectations for educators and schools. They expect schools to

- Provide access to opportunity for *every* student to develop his or her full potential
- Differentiate instruction to meet all students' needs, with extra help for those who have fallen behind and special enrichment for students who are already advanced beyond their peers
- Transmit the legacies of multiple cultures
- Cover a required curriculum that encompasses vast amounts of information

- Prepare all students to pass standardized tests (whose items may not correlate with the endorsed curriculum standards)
- Develop students' ability to think critically, apply their knowledge in the world, and be continuing learners
- Enable students to succeed in the modern workplace[1]
- Produce citizens who will sustain local and global communities

All these stakeholders ply the schools with often incoherent and incompatible combinations of requirements, resources, constraints, and complaints.

Addressing these multiple agendas has always challenged and frustrated educators. During periods of rapid change, like the present, both the imperatives and the difficulties for providing excellent public education become even more intense. Astonishing technological developments during the past century have transformed the nature of knowledge and work, the speed of travel and communication, modes of warfare, and humanity's impact on the planet. All of these conditions make access to high-quality learning essential for all people. Economic development, international peace, democratic government, and preservation of the global environment all depend on universal education. Furthermore, traditional basic education in the three R's of reading, writing, and 'rithmetic is nowhere near adequate for modern conditions. Nor is traditional instruction, with teachers leading rows of students through standard curriculum materials, sufficient to prepare today's students to succeed as stewards of tomorrow's world. Teachers must be supported in their efforts to teach new curricula, using new methods and new technologies, to a newly diverse range of students.

In the face of competing agendas and mounting pressures, how can teachers make responsible choices as they design lessons and orchestrate learning with their students? How should school administrators establish priorities, design structures, and foster an organizational culture that galvanizes all members of a school community in a focused and sustained pursuit of coherent goals? How might policymakers, business leaders, and supporters of public education make contributions that truly help school people meet the challenges and serve the purposes of public education in these complex times? How can new technologies support these processes? Educators do not agree on the answers to these questions. They don't even share a common language for articulating their own answers or forging them into a coherent plan.

What This Book Is About

This book outlines a coherent process to help educators address these questions. The Teaching for Understanding framework is distilled from years of analysis by researchers at the Harvard Graduate School of Education, in collaboration with reflective schoolteachers working in a range of subject matters and settings. The elements of the framework guide deliberate decisions about fundamental issues that all educators must resolve. These include framing curriculum topics, defining educational goals, designing learning activities, integrating coherent and effective assessments, developing supportive learning communities, and integrating new technologies to improve learning.

The framework does not supply particular answers, but it helps decision makers involved in all sorts of educational contexts develop effective strategies for their own particular circumstances; it is applicable for all subject matters, ages of learners, and types of settings. Classroom teachers, principals, curriculum planners, technology specialists, policymakers, teacher educators, university professors, corporate trainers, business leaders, researchers, parents, and philanthropists can apply it to their work. In so doing, they learn how to clarify, negotiate, and coordinate their efforts in ways that integrate new educational technologies to advance teaching and learning for understanding.

The book is organized to help a broad range of readers understand the underlying principles behind this model and learn to use it to help guide their work. Chapter One introduces the five elements of the Teaching for Understanding framework: (1) generative curriculum topics, (2) understanding goals, (3) performances of understanding, (4) ongoing assessment, and (5) collaborative, reflective learning communities. It also describes how these elements fit together as a coherent guide for educators.

Chapter Two explains how this framework and new educational technologies can be effectively coordinated to improve student performance around key learning goals.

Each of the next five chapters concentrates on one element of the framework, elaborating on the key features of the element and illustrating how teachers can use it to guide their integration of new technologies and promote students' learning. The case studies are drawn from one author's extensive experience as a classroom teacher and from all the authors' consultation with a wide range of educators over many years. The examples reveal how teachers make choices as they plan curricula that integrate new technologies, assemble the materials and assistance they need, manage their classroom activities, and develop relationships with colleagues and collaborators within and beyond their school. Most of the examples deal with students in K–12 classrooms, but some vignettes illustrate how this approach works with other learners, including postsecondary students and teachers themselves. In all the examples, the central goal is to advance learners' understanding and their ability to apply knowledge in their lives.

Chapter Eight describes how educators learn to apply the framework to their own practice. It illustrates how the framework itself can be explicitly applied and modeled in designing professional development to support teaching for understanding with new technologies. The examples in this chapter are based on the authors' experience teaching graduate courses and leading professional development for educators, both in on-line courses and in on-site sessions.

The urgent rationale for taking action and some guidance for doing so are summarized in Chapter Nine. Thoughtfully integrating new technologies to help students develop flexible and creative understanding is a complex process that requires sustained inquiry and collaboration among professional communities of educators and between schools and their surrounding communities. This ongoing process depends on coordinated support from policymakers, educational administrators, professional developers, and teacher educators.

Why This Book Is Needed Now

People in schools, as well as those who hope to support them, must respond to the urgent need for improved public education. All young people must now be educated in ways that enable them to be responsible citizens of the planet and of their local communities—a new kind of citizenship. The times generate crushing pressures on educators, with broad mandates to serve increasingly diverse student populations, narrowly focused accountability measures, and often chaotic administrative contexts in which contradictory policies generate competition for resources. Yet these times also provide new opportunities, including an emerging consensus about the nature of effective learning and increasing access to new technologies that support such learning. Teachers must be helped to perform in ways that go far beyond traditional conceptions of delivering instruction. Educational technologies can be used to promote the kinds of learning and teaching that today's world requires. These goals are challenging, but they must be accomplished.

The Authors' Ongoing Collaborative Inquiry

For the past two decades, the authors have worked to improve public education with new technologies in numerous ways—as researchers, professional developers, teacher educators, technology-integration specialists, and classroom teachers. Firsthand experience has revealed both the challenges and the benefits of transforming educational practice from an emphasis on transmitting knowledge to constructing understanding, from isolation and competition to interaction and collaboration, and from trivial or distracting uses of technologies to applications that profoundly enrich and extend learning. Experience has also demonstrated the value of the Teaching for Understanding framework for guiding the thoughtful use of new technologies to support effective education.

The authors' understanding has evolved over decades of educational research, conducted both independently and during several periods of collaboration. Stone

Wiske has worked for over twenty years at the Harvard Graduate School of Education. At Harvard's Educational Technology Center, she directed collaborative research about ways to use technologies to improve the teaching and learning of important topics—sometimes called targets of difficulty—in mathematics, computing, and science. She also worked with colleagues from Harvard's Project Zero for over ten years on a project that developed the Teaching for Understanding framework that forms much of the conceptual foundation underlying this book. Wiske conducts research and teaches graduate courses about educational design that integrates new technologies and focuses on teaching for understanding. Since the mid-1990s, she has developed and studied ways of using networked technologies to connect educational research with practice. This work includes a professional development program that uses the Internet to engage educators in interactive professional communities of inquiry aimed at improving public education.

Kristi Rennebohm Franz began teaching in 1989, with a commitment to creating classrooms in which even primary-grade school children would learn to become responsible and productive citizens by connecting their schoolwork to authentic problems in the local and global community. Her goals have always encompassed not only ensuring that her students learn *what* they need to know but also *how* they develop enduring understanding that they can apply appropriately in the world and *why* this is important.

Beginning in the 1993–94 academic year, Franz started integrating new technologies into her classroom to promote more effective collaboration between her classroom and other teachers, students, and resources in the world. She was among the first elementary teachers to join a global education network called iEARN (the International Education and Resource Network), through which her classes participated in a number of collaborative projects, with remarkable social and civic results.

Franz was a visiting scholar at Harvard in 1997; her goal was to clarify and articulate how new technologies can provide profound educational leverage in her classroom. She knew that her young students were making academic progress and learning how to be responsible global citizens in ways that far surpassed most people's expectations for K–2 students. She believed that new technologies were partly responsible, but she lacked a framework and a group of colleagues to help formulate and test her hypotheses. Wiske and Rennebohm Franz found that the Teaching for Understanding framework provided a set of concepts and a common language for mapping connections among their goals, practices, and insights.

Lisa Breit directed a multiyear program to integrate educational technologies into teachers' practice in the Watertown, Massachusetts, schools. Beginning in 2002, her doctoral work at the Harvard Graduate School of Education focused on teachers as instructional leaders, with an emphasis on the use of new technologies to support both teachers' learning and their work with students.

Through many conversations and much collaborative teaching, the three authors refined their conception of the Teaching for Understanding framework, designed and described classroom practices that specifically highlight elements of the framework, and synthesized recommendations for putting these approaches into practice. They have applied these ideas with a wide range of learners, including school children in all grades, graduate students, beginning and veteran teachers, teacher educators, and educational leaders and policymakers. These experiences demonstrate that collaboration and conversation structured by a shared language based on an explicit educational model help educators design and refine ways of using new technologies to improve teaching for understanding. Changing teaching practice is a gradual process that must be supported through cycles of trying an innovative approach, analyzing how and why it worked, devising refinements, and working through the cycle again. All of this is difficult to accomplish alone. It benefits from just the sort of reflective, collaborative dialogue with like-minded colleagues that has generated this book.

This book can help educators promote cycles of innovation, dialogue, and inquiry toward teaching for understanding. Such inquiry is endangered by the current context of competing political priorities and intense pressures to hold teachers and students accountable using overly narrow measures of student performance. Professional inquiry is threatened by a welter of incoherent mandates and by autocratic insistence on standardized procedures for achieving limited objectives. We hope to support professional collaboration that is coherent, that generates clear evidence of improvement in teaching practice and student performance, and that engages teachers and other educators in nourishing and fruitful inquiry.

How to Use This Book

Instead of prescribing specific answers to key questions about education, the framework presented here provides a clear, cogent, and practical structure to help educators make their own decisions. The framework helps educators address mandated curriculum and assessment priorities while honoring their own expertise. The tone and form of the book reflect the authors' dual commitment to research-based principles and time-tested practices. The combination of clear criteria and vivid cases can help educators see how to apply this framework in their own situations, taking into account their particular preferences, priorities, learners, and circumstances. Although most of the examples are drawn from schools, this approach is applicable in all kinds of educational settings, both formal and informal, with any subject matter and for all kinds of learners.

The goal of this book is not simply to present a set of useful concepts but to stimulate educators to apply these ideas in ways that improve teaching and learning, thus generating both success and satisfaction for teachers and their students.

Effective designs for learning do not result from replicating "best practices" or dutifully following a recipe. They always require a degree of invention, interpretation, and adaptation by thoughtful teachers attuned to the particular demands and opportunities of their own situation.

So why offer people a book, knowing that reading about ideas is not sufficient to enable educators to accomplish ongoing inquiry and significant change in daily practice? The hope is that readers will use the book while participating in a community of other people who are interested in collaborating with them to think and work with these ideas. Such groups might be formed within schools or professional associations, at universities, or through collaborations across organizations. The book may provide useful guidance for teacher study groups,[2] leaders of staff development and curriculum design, coordinators of educational technology, teacher educators, and others engaged in systematic school improvement.

Ideally, as readers work through this book, they will take time to try out new strategies, such as drafting policies, revising curriculum, designing lesson plans, or applying new pedagogical strategies. Under such circumstances, the book can become a working guide by which to structure ongoing collegial reflection and dialogue on questions about teaching: How might I apply this idea in my situation? What did you try? How did it go? What might we do next? In light of this experience, now what do you think about this framework?

Additional resources and structures are available to help educators apply the approach outlined in this book. The case studies and vignettes throughout the book refer to a range of print materials and Web sites that teachers may find helpful. Chapter Eight includes extensive descriptions of interactive on-line resources and professional development activities designed specifically to support teaching for understanding with new technologies. Readers who might want to use these resources to design curriculum or take on-line courses about these ideas may want to read Chapter Eight first.

In all these endeavors, a continual process of considering new educational ideas and principles, translating them into terms that make sense in one's own circumstances, developing a plan based on this thinking, trying out the new approach, and reflecting on the results is essential. Each of these steps benefits from collaboration and dialogue with colleagues who share common goals and a common language for discussing teaching practice. We hope this book stimulates and supports collaborative, reflective communities of educators committed to ongoing inquiry and invention.

Cambridge, Massachusetts   *Stone Wiske*
May 2004

NOTES

1. For a recent summary of what business leaders, policymakers, and educators think students need to know to succeed, see *Learning for the 21st Century,* published in 2003 by the Partnership for 21st Century Skills, downloaded from http://www.21stcenturyskills.org/downloads/P21_Report.pdf on April 26, 2004.

2. The process of lesson study, as carried out by teachers in Japanese schools, would be an excellent structure for making sense of this book. For a description of this process, see *The Teaching Gap: Best Ideas from the World's Teachers for Improving Education in the Classroom* (Stigler, J. W., and Hiebert, J. New York: Free Press, 1999).

Acknowledgments

This book synthesizes more than twenty years of learning that was spurred on by too many colleagues to thank individually. So I will mention only those whose ideas were especially influential in shaping my understanding.

At the Educational Technology Center at the Harvard Graduate School of Education, Judah Schwartz, Greg Jackson, and Charles Thompson first introduced me to ways of thinking about the potential of new technologies to improve teaching and learning of important "targets of difficulty." David Perkins also shaped my thinking in those days and taught by example how to develop ideas in groups. He, along with Howard Gardner and Vito Perrone, were the guiding minds behind the Teaching for Understanding project. They and other colleagues, especially Lois Hetland and Joan Sobel, were especially important in stimulating my understanding of the framework developed by the Teaching for Understanding project through years of collaboration among teachers and researchers.

Connecting the ideas developed at the Educational Technology Center with those formulated through the Teaching for Understanding project is an ongoing endeavor, which I have pursued for the past decade. During that time, Jim Moore, Donily Corr, Susan Wirsig, Mindy Sick Munger, Shannon Martin Croft, Cheryl Campbell, and Heidi Soule have been especially thoughtful and generous in helping me see how to make sense of these ideas and to apply them effectively in teaching. Using on-line technologies to support this work has been my focus during the past five years at the WIDE World (Wide-scale Interactive Development for Educators) project with David Perkins and Nathan Finch, among many other remarkably creative and dedicated colleagues.

Integrating these strands of work into a clear, vivid, and useful form is the goal of this book. Kristi Rennebohm Franz has been a powerhouse of inspiration, support, and encouragement for this process. Her brilliance as a teacher is captivating, her determination to analyze good teaching is incessant, and her generosity as a collaborator is unbounded. We have discussed all the ideas in this book; many of them originated with her, and she has read every page through several drafts. Through all this dialogue, Kristi's good spirit, excellent mind, and deep devotion to humanity supplied both momentum and meticulous care. Kristi provided all the material for the case studies of her practice in Pullman, Washington. Lisa Breit, another bountifully thoughtful educator, joined our conversation during the past two years and has added sparkle, breadth of experience, and her own special enthusiasm to the endeavor. Lisa wrote the vignettes that portray teaching for understanding with new technologies.

Many other creative teachers have shared their ideas and experiences with practical examples that illustrate the themes of this book. They include Marta Liebedinsky, Cesar Nunes, Mary Teixiera, Lisa Martinez, Scott Weatherford, Audrey Ting, Miranda Whitmore, Janet Jehle, Patricia Norris, Amy Fritz, Kate Paterson, Beth-Ann Keane, Linnie Regan, Monica Hiller, Lisa McDonagh, Ellen Fitanidies, Katie Beller, Allison Levit, Linda Picceri, Wendy Hankins, and Bernie Dodge.

Several especially generous colleagues read an early draft of this book and offered astute advice about how to improve it: Robert Reich, Milton Chen, Margaret Riel, Les Foltos, Adriana Villela, Ed Gragert, Mary Skipper, and Kathy Klock. Lesley Iura, my editor at Jossey-Bass, has been a steady source of good suggestions and encouragement about this work. Linda Chisom has provided administrative support with unfailing good cheer and sound judgment. I have tried to heed the excellent recommendations of these wise readers, but any remaining mistakes or flaws are my responsibility. To all of these teachers and collaborators, I extend my deep appreciation and my hope that you will find your own good work carried along through these pages.

The Authors

Martha Stone Wiske is a lecturer at the Harvard Graduate School of Education where she codirected the Educational Technology Center from 1997–2003. She teaches, conducts collaborative research, and consults with educators both nationally and internationally, with a focus on the integration of new technology to enhance teaching and learning. Her projects include the Education with New Technologies Web site at http://learnweb.harvard.edu/ent and an on-line professional development program called WIDE World (Wide-scale Interactive Development for Educators at http://wideworld.pz.harvard.edu). WIDE World applies educational research to improve teaching practice through a range of on-line programs and on-site professional development activities for educators.

Wiske edited *Teaching for Understanding: Linking Research with Practice* (Jossey-Bass, 1998). She is cofounder of ECi (Education, Communication, and Information at http://www.open.ac.uk/eci), a journal for international dialogue about new developments in educational theory, practice, and technology.

Kristi Rennebohm Franz, a Washington State teacher, has designed and implemented curricula with new technologies that connect school communities worldwide. Her classroom teaching and contributions to education have been recognized with the Peace Corps's Paul D. Coverdell World Wise Schools Excellence in Education Award, the Presidential Award for Excellence in Teaching Science, the Milken Educator Award, and the Washington State University Dr. Martin Luther King Jr. "Keeping the Dream Alive" Award, a Scholastic Blue Ribbon Website

Award, and by the International Reading Association. Her classroom teaching has been filmed and featured in the PBS documentary "Digital Divide" and the North Central Regional Educational Laboratory's enGuage Professional Development Program. She has authored invited articles and chapters for publication, including the *Theory into Practice* journal (Ohio State University College of Education, 1996) and the book *Teaching and Learning in the Digital Classroom* (Harvard Education Press, 2003). She is on the editorial board for the international journal, *Education, Communication and Information* (ECi). She has provided models of education and articles for the U.S. Department of Education; her pedagogy has informed education action in the U.S. Senate. She had Visiting Scholar and Visiting Practitioner appointments at Harvard's Graduate School of Education in 1997 and 2002. She has been a Lead Teacher for the International Education and Resource Network since 1994 and is currently an iEARN project director for International Education Teacher Professional Development. She lives in Seattle, Washington.

Lisa Breit is the former director of technology integration for the Watertown, Massachusetts, public schools, where she managed a technology professional development and research initiative called MetroLINC. She devised and led the district's comprehensive teacher development program for technology integration, including workshops, institutes, and direct consultation with teachers as they designed and implemented curriculum with new technologies over a six-year period. She also advised the district on how to build infrastructure, cultivate leadership, and provide institutional support as teachers gained proficiency with instructional technology.

Breit is a member of the adjunct faculty for the Lesley University's Technology in Education master's program and has presented numerous workshops at conferences, including MassCUE (Massachusetts Computer Using Educators), National Educational Computing Conference, and Improving America's Schools. Currently, she is a doctoral student at the Harvard Graduate School of Education, where she is studying how technology can help improve instruction and support continuous learning for both teachers and students. She holds a B.S. in human development from Cornell University and an Ed.M. from the Harvard Graduate School of Education.

Teaching for Understanding with Technology

ONE

Overview of Teaching
for Understanding with Technology

What Is Teaching for Understanding?

Figuring out how to take full advantage of new educational technologies is a complex process that goes beyond purchasing hardware and plugging it into a power source. And it is not accomplished by merely selecting appropriate software. Educational technologies are not like appliances that automatically do their jobs when the "power" button is pushed. Information and communication technologies such as calculators and computers, as well as networked technologies like e-mail and the World Wide Web, are interactive, rapidly evolving media with which to think and learn. They help to create collaborative social contexts for learning in ways not previously possible. As a result, effectively integrating new technology into educational practice is not just a matter of learning how to use the technology. It is also a process of reflecting on how technology-enhanced practices challenge assumptions about what and how to teach and how students can learn most effectively in today's world.

Underestimating the complexity of this process and failing to support it adequately seems an almost universal shortcoming. Decision makers in school systems tend to buy hardware first, then make choices about software; only gradually do they realize that they must also help teachers learn how to use these new resources before the technology can significantly contribute to students' educational experience. Only after these processes are in motion do educators, policymakers, parents, and other stakeholders usually recognize that they must connect their decisions about educational technologies to their priorities for education.[1]

If new technologies are going to lead to significant improvements in teaching and learning, the process of technology integration must be understood and undertaken as an *educational* process. Decisions about hardware, software, distribution of resources, curriculum design, and professional development should all be based on clear and explicit answers to fundamental educational questions: What should students come to understand? How can learning be promoted and assessed? What role should technologies play in these matters?

This chapter presents a systematic framework to guide the development of answers to these questions. The framework grew out of a sustained collaborative research project conducted from 1991 to 1997 by researchers at the Harvard Graduate School of Education, along with groups of effective teachers working in a range of subject matters and school contexts. The purpose of this project was to clarify the nature of understanding and then to define features of educational practices that helped students develop deep and flexible understanding. The Teaching for Understanding framework emerged as the researchers and teachers analyzed case studies of effective teaching practices in relation to current theories of cognition and instruction.

In the years since the Teaching for Understanding project published its findings and framework,[2] this educational model has served as a structure for designing educational materials and activities in a wide variety of locations and types of settings throughout the United States and around the world: preschools, elementary and secondary schools, universities, and professional development programs. It has guided the design of curriculum and pedagogy in a range of educational initiatives, including efforts to improve teaching and learning through the integration of technology.[3] The framework has proved to be an accessible model that is "roomy" enough to encourage professional judgment yet specific enough to guide educators' progressive refinement of their work toward promoting effective understanding. Because of these qualities, it provides generally useful guidelines for designing education, as well as a framework with which to focus the integration of new technologies on learning.

The framework has proved to be an accessible model that is "roomy" enough to encourage professional judgment yet specific enough to guide educators' progressive refinement of their work toward promoting effective understanding.

What Is "Understanding"?

Any discussion of educational plans must be grounded in a conception of the ends or purposes of education. Schools have been expected to serve a range of purposes historically: cultural assimilation, civic preparation, economic development, academic achievement, and individual fulfillment. Currently, schools are also pressed to update their practices and to prepare both teachers and students for the twenty-first century, taking account of developments in new technologies and trends in global politics, economics, and cultural interactions. In the context of such complex and evolving goals, formulating a clear and compelling, yet flexible, conception of the purpose of education may seem impossible.

"Understanding," as characterized by the Teaching for Understanding project, appears to provide a workable, specific yet generally applicable articulation of what schools ought to help students learn. After reviewing both educational research and the effective practices of teachers, the project defined *understanding a topic* as being "able to perform flexibly with the topic—to explain, justify, extrapolate, relate, and apply in ways that go beyond knowledge and routine skill. Understanding is a matter of being able to think and act flexibly with what you know."[4]

Understanding as a "flexible performance capability" encompasses four dimensions: (1) *knowledge* of important concepts, (2) *methods* of disciplined reasoning and inquiry, (3) *purposes* and limitations of different domains of understanding, and (4) *forms* of expressing understanding for particular audiences.[5] (See Chapter Four for examples.) This definition of understanding takes account of research showing that learning is an active process, not simply a matter of absorbing information or practicing basic skills. Demonstrations of understanding-as-performance require the learner to generate products or performances that go beyond reproducing received knowledge. At the same time, this conception of understanding honors the importance of mastering certain bodies of knowledge and methods of disciplined inquiry in domains such as history, mathematics, science, and language.

Defining the goal of education as a flexible capability to think and apply one's knowledge carries implications for the process of learning and teaching. If understanding is *demonstrated* by performance, it follows that understanding is also *developed* by performances of understanding. Such performances require learners to stretch their minds, to think using what they have learned, and to apply their knowledge creatively and appropriately in a range of circumstances. The Teaching for Understanding project made "performances of understanding" the centerpiece of its framework.

What Is "Teaching for Understanding"?

Having defined *understanding* as a flexible performance capability, members of the Teaching for Understanding project proceeded to examine pedagogical practices that foster this kind of understanding in students. Their study included reviewing current research on learning and teaching, as well as analyzing examples of practices conducted by teachers of various subjects in middle and secondary schools.

Through multiple cycles of collaborative research that included writing case studies about particularly effective curriculum units and analyzing them in relation to theories of cognition and instruction, the project defined a model with four elements (listed in the next sections) that incorporate the characteristics of particularly effective teaching for understanding; these elements also help teachers design lessons by formulating answers for some basic questions that all educators must address:

- What topics are worth understanding?
- What exactly should students understand about such topics?
- How will students develop and demonstrate understanding?
- How will students and teachers assess understanding?

What Topics Are Worth Understanding?

From the panoply of possible topics encompassed by curriculum standards and required or recommended textbooks, how should teachers decide what to teach? Teaching for understanding requires that students make sense of what they learn, not just memorize facts and formulas. Therefore, curriculum should be organized around topics that are meaningful to students, as well as important to the subject matter. If understanding includes a capacity to think with what you know, it follows that curriculum topics should not simply be "covered" but "uncovered" in ways that invite continuing inquiry. Teachers are best able to guide inquiry around topics that they themselves find endlessly fascinating.

With all these factors in mind, the Teaching for Understanding project recommended that teachers organize curriculum around *generative topics* that have the four features mentioned earlier. The topics should be (1) connected to multiple important ideas within and across subject matters, (2) authentic, accessible, and interesting to students, (3) fascinating and compelling for the teacher, and (4) approachable through a variety of entry points and a range of available curriculum materials and technologies. Generative topics have a "bottomless" quality that generates and rewards continuing inquiry.

What Exactly Should Students Understand About These Topics?

Even very thoughtful and conscientious educators often struggle to define clearly and exactly just what they hope their students will come to understand. As long as these goals remain tacit, perhaps even for teachers themselves, students will be uncertain about what they should be striving to accomplish.

Teachers and students are better able to concentrate their efforts when understanding goals are clearly defined and publicly stated. *Understanding goals* should focus on big ideas that go beyond memorizing facts and rehearsing routine skills. They may address multiple dimensions of generative topics, including key concepts, disciplined modes of reasoning, underlying purposes for learning, and mastery of forms for expressing learning. Goals for a particular lesson are coherently connected to larger goals for a longer curriculum unit; unit goals, in turn, clearly connect to even larger term-long or year-long

A clear and coherent, nested set of goals helps both students and teachers focus on the core purposes of every aspect of the learning process.

goals. The Teaching for Understanding project referred to such overarching goals as "throughlines," because they serve like an actor's throughline to shape and focus a whole strand of performances. A clear and coherent, nested set of goals helps both students and teachers focus on the core purposes of every aspect of the learning process.

In keeping with the emphasis on understanding as performance, understanding goals clarify what students will be able to do with their knowledge. Statements of goals usually include action verbs like *appreciate, analyze,* and *explain* rather than more passive phrasings such as *know that, list,* or *correctly use.* Understanding goals may require students to learn particular facts and to develop skills, but they also require students to think with this knowledge and apply it in creative ways.

How Will Students Develop and Demonstrate Understanding?

Performances of understanding are the means of developing and demonstrating understanding. They are the centerpiece of the Teaching for Understanding framework, and they should constitute a large portion of the work that students do. Such performances build students' understanding of target goals and involve learners in activities that require creative thinking. In order to engage learners in such performances, teachers need to design a sequence of activities that start with introductory activities that build on students' beginning interests and knowledge. Through a series of guided performances, teachers help students gradually acquire new knowledge, along with the ability to apply their knowl-

Understanding goals may require students to learn particular facts and to develop skills, but they also require students to think with this knowledge and apply it in creative ways.

edge in creating increasingly sophisticated products and performances. Students should then be able to work more independently of the teacher in producing a culminating performance that synthesizes multiple dimensions of their understanding. Effective teachers devise a range of performances that allow students to develop and apply different kinds of intelligences and modes of expression. Of course, some learning activities entail taking in new information from reading or presentations. Teaching for understanding requires, however, that students also engage frequently in activities that require them to think, not just memorize or practice routine skills.

How Will Students and Teachers Assess Learning?

Performances that develop and demonstrate students' understanding also provide an occasion for assessment. And the Teaching for Understanding project found that effective teachers provide *ongoing assessment,* along with coaching, to help students gradually refine and improve performances of understanding. Unlike traditional assessments that are conducted at the end of a course to determine how well a student performed, ongoing assessments are conducted frequently throughout the process of learning. Their purpose is not only to gauge achievement but also to promote better performance by providing specific information about strengths, as well as suggestions for improvement.

In the Teaching for Understanding model, ongoing assessments are based on explicit criteria that relate directly to understanding goals. These criteria (or rubrics) are publicly shared early in the process of developing a performance so that students understand what they are trying to achieve. Indeed, teachers may involve students in developing assessment criteria by analyzing examples of student products or performances and defining the qualities of good work. When assessment criteria are clear and public, learners can use them as a performance guide while they draft and revise their work. Students themselves can also contribute to ongoing assessments of their own performances and of their fellow learners' work. Assessments may be informally embedded into learning activities or conducted through more formal procedures. In these ways, ongoing assessments serve as a means of promoting, as well as monitoring, student performances.

How Will Students and Teachers Learn Together?

Social interaction and reflective conversation within communities of learners have long been recognized as important aspects of developing understanding.[6] The original framework published by the Teaching for Understanding project alluded to this aspect (with its endorsement of peer exchange around performances and assessment) but did not include it as a separate element. Because communities of learners are central for developing understanding and because new technologies provide such powerful tools for supporting them, this element of the framework is developed explicitly in this book and termed *reflective, collaborative communities.*

Community-based learning takes place in a social context in which learners interact with other people who collaborate on work that is meaningful to the group. A community of learners engages in dialogue, using a shared language that helps participants analyze their practice, reflect on their learning, and devise ways of improving understanding. These exchanges depend on the development of norms that include respect, reciprocity, and commitment to cooperation on communal accomplishments, not just on the advancement or improvement of individual performances.

How Do Educators Use This Framework?

The elements of the Teaching for Understanding framework and their key features, as summarized in Exhibit 1.1, may appear to be a static and sequential list.

But in practice, they serve as guidelines and reminders for thoughtful educators as they plan curriculum, teach and assess their students, and reflect on their practice to devise improvements.

Frame the Process of Professional Inquiry

Teaching for understanding is a continual process, not a method that teachers perfect and implement once and for all. The process is part of the ongoing inquiry that professionals carry out as they focus on research-based principles of good practice, apply these principles to design or modify their own practice, and study the results of these efforts to make further improvements.

This recurring cycle of learning, revising existing practices, trying out new approaches, and reflecting on experience is very difficult to accomplish in isolation for several reasons. First, applying general guidelines to the design of curriculum plans and pedagogical practices requires creativity and invention. The Teaching for Understanding framework is not a complete set of curriculum materials. It can be used to analyze or develop plans and materials to teach any topic, in any subject matter. That generic quality of the framework is a strength in terms of its broad applicability, but it also means that teachers must build a bridge from these general principles to their own particular situation.

Second, making time for reflection and invention is difficult in most teachers' lives. Teachers are immersed in practical, concrete decisions, and they have few opportunities to reflect on their subject matter or their pedagogical principles in terms of abstract ideas. They are surrounded by textbooks, required tests, myriad administrative forms to complete, and a continual parade of varied students who absorb all their energy. Analyzing curriculum goals in terms of big ideas that anchor understanding goals is not a usual part of most teachers' day or week. For many teachers, taking this kind of perspective on their subject matter or domain is not something they have experienced since their own college days—or ever.

Just as taking time for reflection is difficult, shifting accustomed practice is also a challenge.

Just as taking time for reflection is difficult, shifting accustomed practice is also a challenge. Teachers assemble a repertoire of lesson plans and classroom strategies that they hone through experience and sustain through habit. In altering their routines, they risk losing control of a precarious process of engaging students in classroom work. Even if a new practice might work much better, that future is hard to anticipate. Letting go of current practice feels risky.

EXHIBIT 1.1. Key Features of the Teaching for Understanding Framework

Generative Topics

- Are connected to multiple ideas within and across subject matters
- Are authentic, accessible, and interesting to students
- Are fascinating and compelling for the teacher
- Can be approached through a variety of entry points and a range of available curriculum materials and technologies
- Have a "bottomless" quality that generates and rewards continuing inquiry

Understanding Goals

- Are clearly defined and publicly stated
- Focus on big ideas, beyond memorizing facts and rehearsing routine skills
- Address multiple dimensions: knowledge, methods of inquiry and reasoning, purposes for learning, and forms of expression
- Are connected coherently so that lesson-level goals relate to long-term goals and to overarching goals, or throughlines

Performances of Understanding

- Develop and demonstrate understanding of target goals
- Require active learning and creative thinking to stretch learners' minds
- Build understanding through sequenced activities from introductory "messing about" to guided inquiry to culminating performances
- Engage a rich variety of entry points and multiple intelligences

Ongoing Assessment

- Is based on explicit, public criteria directly related to understanding goals
- Is conducted frequently and generates suggestions for improving performance
- Includes informal, embedded assessments, as well as more formal structures and products
- Uses multiple sources: self- and peer assessments, as well as feedback from teachers, coaches, and others

Reflective Collaborative Communities

- Support dialogue and reflection based on shared goals and a common language
- Take into account diverse perspectives
- Promote respect, reciprocity, and collaboration among members of a community on communal accomplishments, as well as individual performances

Coaches Can Help

For all these reasons, teaching for understanding is a process that benefits from collegial exchange and supportive coaching. Talking with other teachers who are thinking with the same framework helps teachers build bridges between the abstract principles and their own experience.

Bridge building can be also be supported through dialogue with an experienced coach who has used the framework to guide practice in a range of situations and who has helped lots of other teachers work with these ideas. Effective coaches model the principles of the Teaching for Understanding framework as they help teachers learn how to apply these same principles; coaches may help teachers see connections between their own current practices and some elements of the framework. Coaches can also provide access to examples of curriculum plans, lesson materials, and stories about how teachers gradually introduced elements of the framework into their practice. Coaches may guide a group of teachers through a process of developing trust, sharing examples of their own work, and developing fluency with a shared language for jointly analyzing and improving their curriculum plans. As teachers plan and try out lessons that reflect the elements of Teaching for Understanding, coaches can provide encouragement and suggestions that model the process of ongoing assessment.

The Approach Is Flexible

Experienced coaches also help teachers appreciate that a framework of this kind is not a recipe with exact instructions to be followed. People who wish to shift their teaching in the directions outlined by the Teaching for Understanding framework may approach this process in a wide range of ways. There is no need to address each element in the order listed in Exhibit 1.1. For example, some teachers find ongoing assessment an appealing idea and see ways to introduce aspects of that element into a lesson before thinking about the others. Other teachers may want to study their curriculum materials and requirements with the aim of defining a generative topic through which they could foster their students' understanding of a large set of understanding goals. There is no correct place to start. And there is no need to begin with a large-scale project that is a dramatic change from one's accustomed approach.

As teachers work with this framework, most find that it gradually leads to wide-ranging and sometimes deep changes in their ideas and their practices. Ultimately, they may transform their curriculum, learning activities, roles, and relationships within the classroom and with people and resources outside the classroom; they may also change their assessment procedures and their own perception of themselves as professional educators. But these large shifts evolve gradually and may begin with only small changes, such as stating goals up-front or inserting more opportunities for students to demonstrate their understanding.

Focus Lessons and Leverage Technology

There are at least two ways to use the Teaching for Understanding framework with new technologies. One is to use the framework to design curriculum that integrates new technologies so that the resulting lesson effectively focuses on developing students' understanding. Another way of using this framework, especially for teachers who are already using technology, is as a means of clarifying just how technology provides significant educational leverage. The elements and key features of the framework provide specific language for noticing how the technology improves learning by making curriculum more generative, by making understanding goals more visible, by enriching learning activities so that they become more like true performances of understanding, and by building in opportunities to enact ongoing assessment and revision. Such precise analyses may help teachers articulate how and why their uses of technology are truly beneficial, enabling them to serve as persuasive leaders. Such analysis may help teachers notice ways to improve a lesson that is already promising to make it even more effective in achieving priority goals.

Both of these ways of thinking about teaching for understanding with new technologies are described more fully in the next chapter. The subsequent five chapters provide even more detailed examples and suggestions to support a process of improving educational practice through integration of new technologies.

Colleagues Can Help

How might readers develop a community of colleagues with whom to discuss and try out these approaches? Perhaps it is possible to assemble a small study group of fellow teachers who are interested in working through this book together—talking about the ideas and supporting one another in trying them out. The book may also serve as a support for professional development workshops or courses and for more formal courses about designing curriculum plans or integrating technology. The WIDE World (Wide-scale Interactive Development for Educators) project (http://wideworld.pz.harvard.edu) at the Harvard Graduate School of Education currently offers on-line professional development courses and other resources that address teaching for understanding and the integration of new technologies. More information about the way this program both reflects and promotes the principles of the Teaching for Understanding model is provided in Chapter Eight.

However readers choose to work with the ideas in this book, they are more likely to understand them by trying to support their own learning with the same principles outlined here. In other words, it's helpful to ask oneself these questions:

- How might teaching for understanding with new technologies be generative for me?

- What is an entry point that makes this topic accessible and relevant in my life?

- What particular aspects of this overall approach do I most want to understand?

- How might I put some of these ideas into practice right away in order to promote my understanding?

- How can I reflect on my understanding in ways that allow me to track my progress and decide on next steps?

- How could I develop a community with whom to undertake this process?

Questions for reflection appear at the end of each chapter as a stimulus for this kind of active engagement with the ideas.

Questions for Reflection

1. How is the Teaching for Understanding framework similar to your current teaching approach, and how is it different?

2. What are your thoughts about how this framework might help you plan to make use of educational technologies to support your students' learning?

NOTES

1. Larry Cuban has written extensively on the history of educational technologies and the reasons they tend to have little effect on schools. His particularly illuminating analysis of this process at elementary, secondary, and postsecondary levels is *Oversold and Underused: Computers in the Classroom* (Cambridge, Mass.: Harvard University Press, 2001).

2. An extensive report on the history, results, and application of the Teaching for Understanding framework is *Teaching for Understanding: Linking Research with Practice* (Wiske, M. S. [ed.], San Francisco: Jossey-Bass, 1998). For a more practical guide in applying this framework to curriculum and instruction, see *The Teaching for Understanding Guide* (Blythe, T. San Francisco: Jossey-Bass, 1998).

3. The "Education with New Technology" Web site at http://learnweb.harvard.edu/ent uses the Teaching for Understanding framework as a structure for a range of on-line resources, interactive tools, and pictures of practice that support the integration of technology in curriculum and instruction.

4. Perkins, D. "What Is Understanding?" In Wiske, M. S. (ed.), *Teaching for Understanding: Linking Research with Practice.* San Francisco: Jossey-Bass, 1998, p. 42.

5. These dimensions of understanding are explained more fully in "What Are the Qualities of Understanding?" (Boix M. V., and Gardner, H. In Wiske, M. S. [ed.], *Teaching for Understanding: Linking Research with Practice.* San Francisco: Jossey-Bass, 1998, pp. 161–196).

6. See, for example, *How People Learn: Brain, Mind, Experience, and School* (National Research Council, Washington, D.C.: National Academy Press, 2000). This synthesis of research on effective learning environments highlights four features: knowledge-centered, learner-centered, assessment-centered, and community-centered.

Using New Technologies to Teach for Understanding

Schools and teachers must prepare today's students for a rapidly changing and interdependent world by teaching them how to think with their knowledge and to apply it flexibly and responsibly. The Teaching for Understanding framework, as outlined in Chapter One, summarizes an educational approach designed to help teachers foster such active and flexible understanding in their students. New technologies offer significant potential for supporting this kind of teaching and learning. At the same time, the framework provides a clear structure for integrating new educational technologies in ways that directly support students' learning. In fact, the Teaching for Understanding framework and new information and communication technologies can serve as mutually reinforcing, synergistic educational innovations. Each can make the other more manageable for teachers and more effective in promoting students' understanding.

How Can Teaching for Understanding Guide the Integration of New Technologies?

What do we mean by a *new technology*? Let us define this term to include any new tools for information and communication beyond the ones traditionally used for teaching and learning. New technologies might include video recorders and players, graphing calculators, computers equipped with any kind of software, digital probes linked to a display device such as a calculator or computer, and the Internet, with its hyperlinked, multimedia Web sites and its e-mail and videoconferencing capabilities. Any resource that can be used to help students wonder, think, analyze,

TIP

What do we mean by a *new technology?* The technology has significant potential to enhance students' understanding and is not yet part of the teacher's repertoire of educational tools.

TIP

Why use new technologies to teach for understanding? In learning to use a tool by doing meaningful work, one comes not simply to develop practical skills but to understand the tool's strengths and limitations.

try to explain, and present their understanding may be considered. The main criteria are that the technology has significant potential to enhance students' understanding and is not yet part of the teacher's repertoire of educational tools.

People in schools and universities are under pressure from many sources—parents, business leaders, policymakers, and would-be benefactors—to "get students on the computer." As a result of this pressure, school systems often purchase new technologies, and teachers often feel compelled to incorporate technology into their classrooms, even without a clear educational agenda in mind. In those situations, using technology can become an end in itself rather than a means to a worthwhile educational purpose.

Time with learners is too precious to spend on activities without a clear and important educational purpose. Even if one's goal is to educate students about technology, that purpose can usually be accomplished most successfully by helping students learn to use the tool in the process of doing worthwhile educational work. Learning to use woodworking tools like a saw, hammer, and plane by using them to build a bench is more stimulating and effective than simply pounding nails as pointless practice. By the same token, learning to use a word processor to draft, review, and revise a meaningful product is more educational than practicing skills in isolation. Students can learn how to cut and paste chunks of text, practice keyboarding, make use of the word processor's capacity to track changes, and experiment with formats to improve readability while working on a writing assignment for one of their courses. In learning to use a tool by doing meaningful work, one comes not simply to develop practical skills but to understand how the tool can serve one's purposes and what its limitations are. The Teaching for Understanding framework guides the design of curriculum that integrates technology to support this kind of authentic learning.

In some schools, the integration of technology into the educational program is built into the organizational structure. Instead of having a separate computer room, a computer teacher, and separate computer classes (which tend to isolate computer use from other schoolwork), some schools now have "educational technology specialists" or "academic computer specialists," as they are sometimes called in universities. Their role is to work with faculty members to help them identify ways of integrating technology with their practice to improve teaching and learning. The technology integration specialists help teachers in all subject matters define goals, identify suitable technologies, and plan ways of connecting the use of technology with their curriculum; then specialists help out in class as students use new tools.

Another way schools help structure technology integration is by arranging schedules so that teachers of different subject matters can work together to design cross-disciplinary projects. At TechBoston Academy—a high school in Boston designed to integrate technology in all aspects of teaching and learning—teachers of different subject matters who teach the same group of students have a common

> ## How Technology Integration Specialists Can Help
>
> • Help teachers define goals
> • Identify suitable technologies
> • Plan ways to connect the use of technology with the curriculum
> • Assist in class as students use new tools

planning period. They use this time to plan one extensive cross-subject project, as well as smaller collaborative initiatives. Students use computers and Web sites to conduct research and present their work on assignments for different classes as part of these projects. The framework presented in this book can serve as a common format to coordinate such collaborative planning efforts and keep them focused on helping students achieve multiple understanding goals.

How might educators proceed to make their educational agenda the guiding force in shaping the selection and integration of new technologies? Whether they are planning to use a particular technology or are exploring possible ways of integrating a range of new tools, they would focus on developing answers to the following fundamental questions to clarify their educational priorities:

- What topics are worth understanding?
- What exactly should students understand about these topics?
- How will learning activities be designed and implemented to help students develop and demonstrate understanding?
- How will students and teachers assess learning?
- How will students and teachers learn together?

The five elements of the Teaching for Understanding framework provide a structure for addressing each of these questions with specific key features for refining those answers (as discussed in Chapter One). The key features that describe *generative topics* help teachers identify or define topics that are worth understanding and worthy of the extra effort involved in integrating new technologies. Clearly stated *understanding goals* define explicitly what students will come to understand about the topic and provide a foundation for guiding the design of learning activities and means of assessing student progress. Taking account of the key features of understanding goals assures teachers that their plans for integrating technology will advance students' mastery of important curriculum objectives. Attending to the key features of *performances of understanding* helps educators define a sequence of learning activities that focus on target goals and that provide a challenging yet accessible progression to build students' achievement. By enacting the key features of *ongoing assessment,* educators build in multiple opportunities to monitor and support students' progress, using assessment criteria that focus on target goals. And finally, the key features of *reflective, collaborative communities* remind teachers to consider ways of engaging learners in interactions with one another and possibly with others to promote their learning.

With this structure, educators can proceed to develop and debate an explicit educational agenda that integrates new technologies and plan designs for achieving it. The agenda might encompass only a short lesson, or it might address work to be carried out over a longer period: a week, a semester, or a full year. Regardless of the timeframe, what is important is to develop a plan that focuses on important learning and that connects educational goals, activities, assessments, and interactions in a coherent and explicit way.

If one's focus is to integrate a particular technology, such as a rich Web-based resource or a piece of software that one's school has purchased, these questions should be addressed within the context of that resource. One might begin with any of the questions, depending on which seems an easy starting point. For example, a social studies teacher might discover the Web site called "Investigating the First Thanksgiving: You are the Historian," published by Plimoth Plantation at http://www.plimoth.org/learn/. After reviewing the site and the activities it provides, the teacher might feel eager to use it with her students. But how will this work connect with her curriculum goals? And how will she be sure that students are really achieving important learning?

Thinking about these five questions and working with the elements of the Teaching for Understanding framework, the teacher might choose first to clarify understanding goals. The Web site defines some possible goals (its developers used the Teaching for Understanding framework to guide their design), but the teacher might also want to review the curriculum standards that she is obligated to address. With these goals clearly in mind, the teacher could refine plans for learning activities and devise some assessments with criteria that reflect required curriculum goals.

The elements of the Teaching for Understanding framework guide a process of planning that keeps the focus on students' understanding. With this priority in mind, plans for using new technologies remain centered on the promotion of important learning. Students and teachers may develop technological literacy, facility in teamwork, and other desirable skills *while also* developing students' understanding of important curricular goals.

How Might New Technologies Enhance Teaching and Learning for Understanding?

When teachers fully integrate the Teaching for Understanding framework into their practice, their curriculum and teaching look quite different from traditional classrooms. Generative topics take account of learners' interests and experience rather than simply follow a standard textbook. Understanding goals focus on key concepts and disciplined ways of thinking, not just on the isolated facts and formulas that form the core of many traditional teaching materials. Performances of understanding engage learners in actively debating, constructing, producing, and presenting their understandings, not just listening to or reciting knowledge created by other people. Ongoing assessments engage learners in critiquing their own products and their fellow learners' work, using explicit criteria and developing suggestions for improving the work. In response to ongoing assessment, students usually make multiple revisions of key products and performances rather than treat their first draft as a final submission. Instead of working in isolation, students regularly communicate and collaborate with one another (and perhaps with people outside their classroom) when teaching for understanding is under way.

Overall, teaching for understanding includes much more active and interactive learning than traditional "transmission" kinds of classroom practices. It requires teachers to shift attention from what they are teaching to what students are learning. Engaging learners in a rich range of performances of understanding is likely to require many more kinds of media, modes of learning, and forms of student products than the usual classroom where the three R's of reading, writing, and 'rithmetic are the norm.

What Does Teaching for Understanding Look Like?

Overall, teaching for understanding includes much more active and interactive learning than is found in traditional classrooms. For example,

- *Generative topics* take account of learners' interests and experience rather than simply following a standard textbook.

- *Understanding goals* focus on key concepts and disciplined ways of thinking, not just the isolated facts and formulas that form the core of many traditional teaching materials.

- *Performances of understanding* engage learners in actively debating, constructing, producing, and presenting their understandings, not just listening to or reciting knowledge created by other people.

- *Ongoing assessments* engage learners in critiquing their own products and their fellow learners' work, using explicit criteria and developing suggestions for improving the work. In response to ongoing assessment, students usually make multiple revisions of key products and performances rather than treat their first draft as a final submission.

- Instead of working in isolation, students regularly communicate and collaborate with one another (and perhaps with people outside their classroom) when teaching for understanding is under way.

How can teachers manage to orchestrate these varied, active, and interactive kinds of learning experiences? New interactive, multimedia, hyperlinked, networked technologies offer myriad possibilities for enacting the key features of each element of the Teaching for Understanding framework, beyond what is possible with traditional materials such as books, paper, and chalkboards.

Generative Curriculum Topics

For example, using real-world data from on-line sources such as the U.S. Bureau of Labor Statistics and the National Aeronautics and Space Administration can make curriculum more generative for students. Interactive multimedia can allow students to approach a topic from more entry points than traditional static textbooks permit. Helping students present their work to an authentic audience via the Web can make schoolwork more relevant and interesting to students.

Understanding Goals

Graphing calculators that instantly relate the graphical and symbolic representations of mathematical expressions can help make understanding goals more accessible to students. Simulations that make abstract concepts, such as friction, visible and that allow students to manipulate abstract concepts can help students comprehend the nature and application of key ideas.

Performances of Understanding

Software that allows learners to create and present their ideas with multiple media (images, text, sound, video, dynamic models) can enrich performances of understanding. Digital media enable students to work together on creating products and integrating multiple forms of expression to convey their ideas. E-mail messages, which seem to combine elements of spoken and written language, may help stu-

---

Why Use New Technologies with the Teaching for Understanding Framework?

Networked technologies, including e-mail and the World Wide Web, offer many ways for learners to reflect on their work and collaborate with fellow learners outside their own classrooms. Numerous on-line educational projects engage students and their teachers in collaborative inquiry and social-action initiatives that help students develop a deeper appreciation for their own and other cultures.

The combination of a guiding framework and a rich toolkit of educational technologies supports continual experimentation, analysis, and improvement in teaching and learning.

dents ramp up their reading and writing skills. Students can use generic software packages, such as spreadsheets, databases, word processors, and programming languages, to help them analyze and present their own ideas in ways that resemble the work of practicing mathematicians, scientists, historians, and writers. New technologies augment students' ways of producing—not just consuming—ideas, which is fundamental to understanding.

For example, Lisa Martinez, who teaches English at TechBoston Academy, involves her students in presenting their ideas to classmates using PowerPoint® slides. She provides a list of themes that are relevant to Shakespeare's *Macbeth,* such as "appearance versus reality." Students select a theme, identify lines in the play that illustrate the theme, explain in their own words why this part of the play relates to the theme, and then look on the Internet to find an image from contemporary culture that illustrates the same theme. Students assemble these components on PowerPoint® slides and present their work to classmates. The entire set of slides is saved as a resource for subsequent classes to use as they study themes in *Macbeth.* Martinez finds that her students, many of whom have not previously experienced much success in their inner-city schools, are more motivated to work on assignments that enable them to learn with multiple media rather than text alone. They appreciate developing valuable technical skills while learning language arts. They also invest more effort in projects they will share with an authentic audience, including classmates, parents, and future students, than in products only the teacher will see.

Ongoing Assessment

Computer tools also facilitate cycles of ongoing assessment in many ways. They capture student work in forms that make the process of analysis and revision much easier than it is with traditional, static tools. The Web allows students and teachers to post their work in places where they can get feedback from a wider range of critics, including authentic audiences who really care about learning from the students' work. Archives of student work, assembled as electronic portfolios, help students, parents, and teachers recognize and support student progress from year to year.

Reflective, Collaborative Communities

Work produced with digital technology can easily be combined with other products and revised. These features help learners work together on combining their individual contributions and co-constructing products. Networked technologies, including e-mail and the Web, offer many ways for learners to reflect on their work and collaborate with fellow learners both inside and outside their own classrooms. Numerous on-line educational projects engage students and their teachers in collaborative inquiry and social-action initiatives that help students develop a deeper appreciation for their own and other cultures.

New Technologies and Teaching for Understanding Are Mutually Beneficial

These examples only begin to suggest ways in which new technologies can be used to enhance each element of the Teaching for Understanding framework. Educators who assess the potential of technology in relation to the elements of effective teaching will see many more ways that interactive, multimedia, networked, and hyperlinked technologies offer resources for conducting active inquiry with students.[1] They may also recognize that some applications of technology are educationally trivial or irrelevant to key goals. The main point is that new technologies offer many advantages beyond the traditional tools of books and chalkboards for teachers who want to involve their learners in developing and demonstrating understanding.

To summarize, teaching for understanding and new educational technologies are mutually supportive, synergistic innovations. The Teaching for Understanding framework can guide the integration of new technologies so that these tools provide significant educational leverage. New technologies help teachers enact the elements of this framework, thus making learning generative, focused, active, reflective, and collaborative in ways that are difficult to achieve with traditional school supplies. Educators who work with both the Teaching for Understanding model and new technologies continually discover ways to improve their practice and refine their educational agenda. The combination of a guiding framework and a rich tool kit of educational technologies supports continual experimentation, analysis, and improvement in teaching and learning.

The remaining chapters include more detailed descriptions of ways to use new technologies to support every element of the Teaching for Understanding model. Each of the next five chapters focuses on one element and includes examples that demonstrate how particular software packages and on-line resources can be integrated into educational activities. They also illustrate the process of gradually incorporating new technologies into a teacher's practice and offer specific recommendations for moving along in this process.

Readers who are eager to begin thinking now about using new technologies may wish to explore a Web site specifically developed to help educators integrate teaching for understanding and using new educational tools. The Education with New Technologies (ENT) Web site at http://learnweb.harvard.edu/ent includes a variety of resources, interactive tools, and means of connecting with other educators who share these interests. Visitors to the site might begin by visiting the ENT Gallery, which displays several pictures of practice that illustrate ways different kinds of educators use the Teaching for Understanding framework as a guide for integrating new technologies into their work. These examples illustrate that the overall approach can be applied with any age learners, with any subject matter, and with a wide variety of technologies and organizational contexts—from primary schools to museums to teacher education colleges.

Questions for Reflection

1. Can you think of a way you might want to apply a new technology in your practice? Perhaps you have already tried such an innovation, and you want to refine the design or clarify the educational rationale behind your approach. Or you may have a colleague who has used computers or other technology in a way that you would like to adapt. You may have used technology in another aspect of your life that you believe has potential for improving your students' learning.

2. How might you integrate this technology in a way that provides significant educational leverage—in other words, that directly supports better understanding of important goals?

NOTE

1. There are numerous books describing ways to use new technologies to enhance effective teaching practices. Among the most thorough, readable, and thoughtful is *Teaching with Technology: Designing Opportunities to Learn* (Norton, P., and Wiburg, K. M. Belmont, Calif.: Wadsworth/Thompson, 2003).

TWO

The Elements of the Teaching for Understanding Framework

Generative Topics and New Technologies

Choosing the topics students will study and deciding how to organize curriculum plans are some of the most difficult decisions a teacher makes. And they are among the most important. Especially when teachers invest the extra effort to integrate new technologies, they want to be sure that the topic merits extra planning and is worthy of students' sustained attention.

What kinds of topics are worth teaching for understanding? Teachers and researchers who worked on the Teaching for Understanding project developed a way to answer this question. They reviewed many curriculum units, across a range of subject matters, that teachers found to be especially effective in developing their students' understanding. Some examples were (1) "the sense of place" in a literature class, (2) classification schemes in a biology class, (3) the Industrial Revolution in a history class, and (4) proportional reasoning in a mathematics class.[1]

Analyzing these cases in relation to theories of effective teaching and learning revealed some common features across the units. In each case, the topic was significant because it related to several important ideas in the subject matter, was easily connected to students' experience and interests,[2] and could be approached in multiple ways through a range of curriculum materials and entry points.[3] Also, the teacher was passionately interested in the topic. Finally, the topic had a bottomless quality because the more students delved, the more they generated new questions to investigate. The Teaching for Understanding project coined the term *generative topics* for curriculum topics that met these criteria as a way of emphasizing that they generate and reward sustained inquiry.

Features of Generative Topics

- Connect to multiple important ideas within and across subject matters
- Are authentic, accessible, and interesting to students
- Are fascinating and compelling for the teacher
- Are approachable through a variety of entry points
- Generate and reward continuing inquiry

TIP

Q: How do you decide which topics and which technologies to choose?

A: Focus on topics that are "targets of difficulty."

New technologies are often particularly appropriate and valuable for enhancing the generative qualities of curriculum topics. The Internet can link the classroom to the students' experiences in the real world by connecting schoolwork to authentic problems, actual data, and outside experts and collaborators. Multimedia technologies, with dynamic images, video, and audio, enrich the usual classroom materials and enable students to learn through a broader range of entry points. Expanding the variety of curriculum materials and means for accessing information enables students to pursue their own interests, ideas, and pathways through an investigation rather than simply follow prescribed steps in a set chapter or workbook. Students are more likely to become engaged in studying a topic if they are able to approach the material in ways that particularly pique their interests and suit their preferred ways of learning.

In some respects, the endless wealth of possibilities opened up by new technologies can pose more of a problem than a solution to the teacher faced with the challenge of selecting curriculum topics. How do teachers decide which topics and which technologies to choose? One response is to focus on the problem spots that tend to recur every year. What topics are perennially difficult for students to learn that are also central to the subject matter and potentially made easier by the use of new technologies? The Educational Technology Center at the Harvard Graduate School of Education recommended such "targets of difficulty"[4] in the curriculum as worthwhile topics on which to focus the use of new technologies. Examples include heat and temperature or weight and density in science, ratios in mathematics, and stereotypes in history and social studies classes. These topics are difficult to understand, centrally important to the subject matter (that is, if students fail to understand these topics well, they will not be able to succeed in further studies of the subject), and likely to be made more understandable through the use of new technologies.

Generative topics that are also targets of difficulty are worth investing the extra effort needed to design and implement curriculum that thoughtfully integrates complex educational technologies. By focusing the application of new technologies on such topics, teachers may alleviate some serious and significant

---

Targets of Difficulty

- Perennially difficult problem spots in teaching and learning
- Central and critical to the subject you teach
- Likely to be understood better through the use of new educational technologies

---

teaching and learning challenges in their current program rather than simply dress up lessons that are already working reasonably well or that are only moderately important.

Case Study: Analyzing Patterns Through Quilt Math

Many teachers and students experience math as the least generative topic in school. Too often, math is taught as a set of right answers that adults know, some smart kids can figure out, and many students can't learn at all. Rarely do students experience mathematics as a useful language they can apply to make sense of the world. As Judah Schwartz, one of the founders of the Educational Technology Center observed, "For the most part, the mathematics we teach in the primary and secondary schools is the mathematics already made by other people. Were we to teach language in that fashion, we would ask the students to learn a play by O'Neill, an essay by Emerson, a short story by Hemingway, but we would never ask them to write prose of their own."[5]

For Kristi Rennebohm Franz, a prime target of difficulty was helping her first- and second-grade students learn how to *use mathematics as a language for analyzing and predicting patterns in the world*. She devised the Quilt Math project (http://www.psd267.wednet.edu/~kfranz/Math/quiltmath2000/quiltmath2000.htm) to make this target of difficulty into a generative topic, and she gradually developed ways of using new technologies to intensify her students' investigations of the mathematics in this project.

Effective Generative Topics

Subject Matter	Generative Topic
Literature	Sense of Place
Biology	Classification Schemes
History	The Industrial Revolution
Social Sciences	Stereotypes
Mathematics	Proportional Reasoning

TIP

For Kristi, a prime target of difficulty was helping her first- and second-grade students learn how to *use mathematics as a language for analyzing and predicting patterns in the world.* Making predictions is key to building generative thinking by giving students an opportunity to create math ideas based on their own analysis of math data and patterns.

On the first day of school, Kristi began Quilt Math by taping a square of fabric (approximately 6 by 6 inches) to the classroom whiteboard and asking her students, "What do you see?" Kristi intentionally selected fabrics for math quilts that related to other curriculum themes the students were studying. For example, at the beginning of the year when students were studying insects, she chose fabrics with ladybug designs. A student might start by saying, "The patch has ladybugs on it," or "The ladybugs are red."

Building on what students said, Kristi asked students to think about how they could add a "math idea" to their comments. Then students might say, "We have one patch with ladybugs on it," and "It is a square patch with nine ladybugs." Thus began a process in which Kristi added one patch per day, five patches per row until five rows generated a 5 by 5-inch patchwork quilt over the first twenty-five days of school. The patches and their patterns became the focus for children as they observed, predicted, described, represented, analyzed, and modified patterns in quantity, shape, and color.

Although students' comments on the first day were few and short, with each added patch the quilt patterns became more complex, and the mathematical thinking expressed by the students grew exponentially in depth and breadth. The class used word processors, digital cameras, math software, and Web site publication tools to enhance the generativity of their mathematical studies. With these tools, they captured and represented students' mathematical ideas in ways that enabled students to analyze patterns in the physical quilt, generate mathematical variations with a virtual quilt, and communicate their ideas on-line in forms that deepened students' understanding. Technology helped students connect important math ideas to their own experience, approach mathematics through a range of entry points and materials, make mathematics authentic and interesting to students and compelling for the teacher, and deepen their understanding through continuing inquiry.

Generative Topic Features in Quilt Math

The class uses of new technologies enhance the generativity of their mathematical studies because they

- *Connect* mathematical ideas to students' own experience
- *Approach* mathematics through *multiple entry points:* visual, symbolic, verbal
- *Deepen* understanding through continuing inquiry about patterns in the physical quilt and the virtual quilt
- *Communicate* ideas on-line in forms that make mathematics authentic and interesting for students and the teacher

Writing Math Comments and Predictions

Students were eager to see what the new patch looked like each morning and to start writing their individual math comments, either on the computers in the classroom or in their paper notebooks. They recorded their observations in preparation for the Quilt Math discussion with the whole class.

During the students' individual writing time, a few students helped Kristi take a digital image of the quilt, download the image to the computer, edit the image, and insert it into a digital math photo journal, where the whole class later recorded its math comments. Students rotated through their turn as helpers with the digital image technology, so everyone in the class learned valuable math lessons while using the software tools for image editing—for example, proportional size cropping, scale dimensions of width and height, measurement in pixels, inches, and percentages, and relative numerical gradients of brightness and contrast. Students learned how to work with the tools for digital imaging by using them for the meaningful purpose of building their Quilt Math journal.

After writing their individual Quilt Math comments, students gathered as a whole class in front of the quilt to share comments. As students described their observations verbally, Kristi recorded their oral comments on the whiteboard, and two students started typing these comments on the computer.[6] The class re-read and discussed the comments to be sure the math reasoning was clear and to support everyone in comprehending the math ideas being expressed by peers. As the quilt grew, the class developed ways of representing key features of the quilt, such as assigning a letter to represent each major color or fabric design in a patch. Then students could characterize the sequence of patches in a row with a formula. For example, a row of patches in the pattern

Yellow, Red, Red, Yellow, Red   became represented as A, B, B, A, B.

Subsequent Quilt Math designs became more complicated with each patch pieced from multiple fabrics. Students started using capital letters for the design or color of the large patch plus numbers for fabrics within the patch. For example, in a star quilt connected to the class astronomy science unit (http://www.psd267.wednet.edu/~kfranz/Math/quiltmath2000/starquilt.html), each square patch in a row was pieced from triangle shapes, with iterations of four different fabrics. Students represented this more complicated pattern as

A(1 + 2), B(3 + 4), C(2 + 3), D(4 + 1)

Such symbolic representations helped the students see how to recognize and communicate about patterns within patches and across rows. In this process, primary students began to learn important lessons about patterns and symbolic representations that paved the way for learning pre-algebra conventions in later grades.

Generative Topic: Feeding the Family: Balanced Diet–Balanced Budget[1]

Students often have difficulty understanding how to apply the math and science they learn in school to the complex contexts they encounter in the real world. To address this target of difficulty, the generative topic of this multidisciplinary unit for sixth-graders focuses on how to plan and shop for a week's worth of nutritious family meals while staying within a budget. Students "shop" for food using the PeaPod on-line grocery database, where they consider nutrition labels, special sales, generic versus brand-name products, coupon discounts, and unit pricing. The site's multimedia interface and twenty-four-hour access allow students to "shop" conveniently and flexibly without costly errors or impractical field trips.

After a series of lessons about nutrition and a guided exploration about unit pricing and family food budgets using interactive Web sites, the students work in pairs to plan a week's worth of meals for a hypothetical family with specific nutritional and financial needs. As students plan shopping lists, they learn to balance nutrition requirements and healthy eating choices with financial constraints.

The PeaPod on-line-shopping Web site offers efficient access to a detailed database of thousands of products organized just as in a typical supermarket, with product images, nutrition labels, and unit pricing information for each item. Students analyze their virtual purchases (substituting products, modifying quantities, and tracking the total cost of their selections) using a spreadsheet to record, organize, compare, and analyze information from the database. Working together, students engage in mathematical reasoning, data analysis, and negotiation, as well as rich discussions about good nutrition on a budget. PeaPod's database of products with sorting features and real-time data, combined with the spreadsheet as an analytic tool, allows students to think about generative connections among key ideas in math, economics, and nutrition within a real situation that is relevant to their own experience.

Resources

PeaPod Online Grocery Shopping Service
http://www.peapod.com

U.S. Department of Agriculture Food and Nutrition Information Center
http://www.nal.usda.gov/fnic/etext/fnic.html

Understanding Nutrition Labels
http://kidshealth.org/kid/stay_healthy/food/labels.html

Using Unit Pricing
http://www.fmi.org/consumer/unit/howtouse.htm

Nutrition Navigator
http://navigator.tufts.edu/

Eat Well for Less
http://oregonstate.edu/dept/ehe/ewfl/

---

[1]This example is based on the work of Linnie Regan.

At the end of each Quilt Math lesson, students made predictions about the patch that would be posted the next day. Students had to explain the basis for their prediction, using math concepts and the patterns that were already evident in the quilt. Different students made various predictions with reference to the multiple mathematical dimensions represented in the quilts—for example, color of patch or number of ladybugs. Making predictions was key to building generative thinking by giving students an opportunity to create math ideas based on their own analysis of math data and patterns. In addition, the students' eagerness to learn what predictions would be confirmed fueled their desire to return to Quilt Math the next day.

The students' eagerness to learn what predictions would be confirmed fueled their desire to return to Quilt Math the next day.

Building the Quilt Math Photo Journal

By the end of each daily Quilt Math discussion, the two students who were the designated word processor recorders for the day had finished copying the summary of comments that Kristi created on the whiteboard. They stored this text file with the digital image documenting the appearance of the quilt, which helped students check the accuracy of the comments. The recorders raised questions of clarification with the class and the teacher to be sure that all the students' ideas were accurately and completely captured. The teacher then helped the recorders edit this document to correct spelling and syntax. This process gave Kristi an opportunity to assess how the writing and reading skills of the two recorders were developing. Each day, the new pages of comments and predictions were printed and added to the Quilt Math photo journal notebook as an archived record of daily lessons. The class could re-read these comments at the beginning of the next day's Quilt Math lesson, or the teacher might assign the pages for reading practice to individuals or pairs of students. The Quilt Math journal was also posted on-line in the classroom Web site so students could revisit their work to review math ideas, make comparisons of changes in the quilt over time, and share the Quilt Math project with their families by accessing the Web site on a computer at home or in the public library.

The Quilt Math journal was also posted on-line in the classroom Web site so students could revisit their work to review math ideas, make comparisons of changes in the quilt over time, and share the Quilt Math project with their families by accessing the Web site on a computer at home or in the public library.

The class also used the Web site to share the Quilt Math lessons on-line with school peers in other locations within the United States and around the world as part of their participation in an International Education and Resource Network (iEARN) curricular project, "Connecting Math to Our World." More about the iEARN project can be found at http://www.orillas.org/math/ and in subsequent chapters of this book. More about Kristi's Quilt Math class is on-line at http://www.orillas.org/math/20012002/justforfun.html.

Key Features of Generative Topics

Although this example describes work with very young children, it illustrates features of generative topics that are important when choosing curriculum topics for learners of any age. This topic addresses ideas that are *central to the domain* of mathematics: analyzing patterns using multiple symbol systems and making conjectures. The topic also addresses a "target of difficulty" in that children perennially have difficulty appreciating mathematics as a process of actively forming and testing conjectures, not just memorizing mathematics created by others. The Quilt Math project was *authentic, accessible, and interesting to students* because quilts were part of their own childhood experiences, math ideas were presented through colorful fabric designs and geometric patterns that held students' aesthetic and visual interest, the fabric related to topics that students were studying at the time (see examples of Quilt Math related to study of Africa at http://www.psd267.wednet.edu/~kfranz/Math/africaquilt/africaquilt.htm and http://www.psd267.wednet.edu/~kfranz/Math/kenyaquilt/kenyaquilt.htm), and the quilts became beautiful and comforting objects in the classroom.

Generative topics are also *fascinating to the teacher.* Kristi became interested in quilts and how they could be connected to math through several women in her town who were avid quilters. They shared her passion for using quilts as teaching tools in the classroom and were able to help her locate the range of fabrics she needed for this project. Kristi's passionate commitment to teaching with quilts increased as she discovered how much students learned about math, writing, new technologies, and other subjects. This project also illustrates a final important criterion for generative topics: they are *connected* to multiple important ideas within and across subject matters and approachable through *a variety of entry points.* The Quilt Math project leads students into appreciating multiple key concepts and modes of reasoning in mathematics, while learning about how to think with several different symbol systems and display their knowledge in a variety of formats. It allows students to engage mathematics through aesthetic images and words, as well as numbers and other symbolic formulas.

How New Technologies Enhance Generative Topics

Kristi believes that students learn by talking and writing about their ideas, recording and sharing these thoughts, and then reviewing their recorded observations to revise and develop new ideas. In the Quilt Math project, her students used new technologies to record both verbal and visual observations to support a process of mathematical inquiry through observation, analysis, conjecture, revision, and communication of ideas.

Connecting Central Mathematics Concepts to Students' Ideas

Students used word processors to harvest their various individual ideas and consolidate them into a consensus document that recorded ideas from the whole class. The text files provided valuable archives of the class discussions about math so that students could see how they developed their math thinking from simple to complex understandings. Students also used this technology to generate representations of their observations in various symbolic systems—an important form of mathematical expression that supports the recognition and prediction of patterns. Once the students captured their comments with a word processor, they could easily edit the document to correct spellings, improve the accuracy of the statements, and interweave additional ideas. In these ways, the word processor helped to make students' own ideas a key component of the mathematics curriculum.

Using Multiple Entry Points

By making a digital photograph of the quilt each day, the class maintained a visual record to correlate with their verbal accounts and mathematical representations in the Quilt Math photo journal. Students used the digital photo journal to review how the quilt looked on any given day and to support in-depth analysis of how the quilt patterns changed over time. Captured in digital form, the Quilt Math journal was easily printed multiple times for students to take home and discuss, and it was uploaded to the class Web site for even wider dissemination. The visual depiction of the quilt was not only a rich additional entry point to complement the verbal descriptions but also a more direct representation of the quilt. The visual representation was a stable record to check the evidence for students' assertions and the accuracy of their predictions about patterns.

Working with multiple representations is an important dimension of mathematical understanding and an overarching goal in Kristi's class. Many students are confused about the roots and role of symbol systems in mathematics. Taking a digital image of the quilt and entering it into a database gave them experience working with one kind of representation of a concrete object. As the students developed ways of representing features of the quilt with letters or numbers, they began to correlate visual, verbal, and symbolic representations of a physical phenomenon.

Kristi's students also used a software program called Shape-Up to create virtual quilts. Using this program, students could replicate the rows, columns, and shapes of the fabric quilt and then change the colors of the quilt shapes to generate new patterns. As they created different versions from the basic quilt layout, they saw multiple iterations of shapes and patterns. The technology enabled students to manipulate the quilt design to generate new math patterns in ways that were not feasible with the fabric patches. In this way, students made their own discoveries about the relationships of lines, spaces, and color in the creation of shapes and patterns.

Connecting Mathematics to the Real World

By posting their Quilt Math photo journal on the class Web site, with its record of their analyses and conjectures along with the visual images, Kristi and her students connected their classroom to the outside world. Students could check the class Web site from home at the end of the day and share their math work with parents. Parents reported that their students used the Web site as a prompt to talk with great detail about the math thinking that happened at school. The class Web site was also a major link between Kristi's class and other classrooms around the world that conducted their own Quilt Math projects. In these ways, the technology helped students appreciate how the mathematics they learned in school was useful for authentic purposes that extended beyond the walls of the school.

Planning to Teach Generative Topics with New Technologies

Figuring out how to organize curriculum around generative topics or targets of difficulty may not be a simple process. Teachers must analyze their subject matter, their learners, the resources that are available, and their own passions. There are a number of ways to begin this process.

Educators might begin by asking, "What do our students always struggle with that is really essential for their success?" That may reveal some targets of difficulty that could be alleviated by incorporating new technologies into lessons. Another strategy is to ask, "What are students fascinated by that might suggest a way into some of the material they need to learn?" Teachers might also formulate generative topics by reflecting on their own interests and expertise, knowing that their capacity and motivation to design inventive lessons will be enhanced by their own passionate engagement. The teachers' own hobbies or areas of deep knowledge may indicate a generative approach to an important curriculum topic.

A professor of architecture figured out a generative topic by building on one of his own passionate interests.[7] He wanted to use new technologies to motivate students who enrolled in his required course on architectural structures—a course that students found notoriously boring because it dealt with physics rather than the aesthetic aspects of architectural design. When asked about how his passions related to structures in architecture, the professor's eyes lit up and he twisted in his chair to pull a book about bridges off his shelf. "I love bridges!" he exclaimed, with eyes twinkling. His imagination was sparked by the prospect of organizing his course around the structure of bridges and developing animated models of bridges to illustrate key structural concepts, such as stress and load.

Mathletics: Understanding Statistics Through Sports[1]

Thinking critically about statistical data is a target of difficulty for most high school students. In this ninth-grade "Mathletics" unit, students engage in analysis and interpretation of statistics through close scrutiny of a topic that is accessible and interesting to them: sports data.

The unit's generative topic is using statistics as they are used in athletics. Students examine how data actually are used in college and professional sports (for example, tracking player and league performance, tournament selections, team finances, and draft picks). They learn how statistics are generated, how to correctly apply multiple uses and interpretations of statistics, how to identify misleading and manipulated data, and how to represent data accurately.

Students begin by researching recent news stories about sports, noting how statistics are used. The teacher then guides the class through a model inquiry about athletes' academic and athletic performance in Big Ten universities. Students make observations and pose questions about the data and note how the data are used in news articles and NCAA policies. As they learn to evaluate claims based on statistics, students develop questions about a sports issue that interests them.

Students then apply mathematics to analyze their topic. They learn to use linear equations, correlation and regression analysis, and a variety of visual representations (plots, tables, charts, and graphs) as they apply statistical methods to the analysis of sports issues. They use graphing calculators and spreadsheets to evaluate sports writers' inferences and predictions, as well as to explore different ways of presenting, analyzing, and displaying sports data. Students use word processors, digital images, and PowerPoint® to record, reflect on, and present their understandings of statistics in sports.

As students delve into the data, they realize that statistics can be accurate or inaccurate, informative, misconstrued, or unclear, depending on how data are processed and represented. They learn how to determine whether claims based on data seem reasonable and are supported by contextual information. Even students who are not typically engaged by math can learn to approach numerical information with a critical eye. By focusing on students' interest in athletics, this unit helps students see how math relates to their world and cultivates interest and skills for ongoing inquiry.

Resources

American Statistical Association K12 Page
http://www.amstat.org/education/index.html#K12

Chance
http://www.dartmouth.edu/~chance/

WWW Resources for Teaching Statistics (including interactive Java Applets)
http://it.stlawu.edu/~rlock/maa51/java.html

ESPN Behind the Numbers (a monthly column on sports statistics)
http://espn.go.com/moresports/sillsarchive.html

Top Sports Sites
http://www.sporthits.com/top_sports_sites/
maintopsports.shtml

The Athletics Statistics Page
http://users.rcn.com/bricklan/athletic/athletic.html

Hyperstat Online (a statistics "textbook")
http://davidmlane.com/hyperstat/

Gallery of Data Visualization: Best and Worst Statistical Graphics
http://www.math.yorku.ca/SCS/Gallery/

iEARN Curricular Project: Connecting Math to Our Lives
http://www.iearn.org/projects/math.html

---

[1]This vignette is based on the work of Audrey Ting.

Tips for Identifying Generative Topics

Begin with these questions:

- What do students always struggle with that is really important to learn?
- What are students fascinated by that might indicate a way into some of the material they need to learn?
- How might I connect my own passions to the curriculum?

Kristi's quilt project grew from her passionate commitment to engaging her students in learning through talking and writing and making predictions. But how could she get them talking and writing about math? The quilt project seemed a good way to make numerical and geometric patterns out of simple materials that would be visually appealing. She was amazed to discover how deeply engaged the students became. Although she had originally planned to do the project only for the first month of school, she continued it all year because it was so successful. Once Kristi realized how much students could learn from talking about patterns in the quilts, she figured out how to build patches to illustrate a range of mathematical concepts: fractions and part-whole relationships, geometric shapes, and algebraic expressions.

What about incorporating new technology—how does that process work? For most teachers, weaving new tools into their lessons is a gradual, somewhat happenstance sequence of events shaped by opportunity and interests. In Kristi's case, she realized that students would think harder about the mathematics in the quilt project if they had a way to record and review their ideas about the quilt. This led her to begin using a word processor to capture the students' ideas. Including a visual record of the quilt's appearance became important as a means of anchoring the conversation about the quilt with more direct, visual evidence. She wanted to encourage her students to talk about math at home, which led her to begin using a digital camera that could readily generate printed images that students could take home. Developing the class Web site, including the on-line Quilt Math photo journal, was one more step in the process of connecting Kristi's classroom mathematics to her students' lives outside school.

Teachers may start by considering the technological resources that they already have or might be able to acquire and asking, "How might these tools help my students learn important material more effectively than they do now?" Are there ways videotape or video cameras, audiotape recorders, computer software, related peripheral equipment such as scientific probeware, or the Internet could make curriculum more generative? Could new tools help make important topics more related to students' interests, more accessible through multiple entry points, more connected to the teacher's own passions? Are there ways of using new technology that would "open up" the curriculum, revealing connections to other key topics and engaging students in more active inquiry?

While pondering these questions, it is important to ask repeatedly, "Is this topic really central to curriculum priorities, and is it something that can't be taught and learned just as easily with less complicated technology?" After all, there is no need to cultivate a pocket garden with a huge tiller when a small spade will do; complicated technologies should be reserved for topics that are both important and difficult.

TIP

A tip for using new technologies is to look for ways of using them that "open up" the curriculum, reveal connections to multiple key topics, and engage students in more active inquiry.

Questions for Reflection

1. What are some topics in my curriculum that are really important yet difficult for students to understand?

2. How might I use a new technology to make one of these topics more generative by connecting it to students' interests and authentic purposes in the world, by building on my own passions and expertise, or by allowing my learners to approach this topic through multiple entry points such as visual art, music, narrative, or mathematical representations?

NOTES

1. For more examples of generative topics, see *Teaching for Understanding: Linking Research with Practice* (Wiske, M. S.) and *The Teaching for Understanding Guide* (Blythe, T. and Associates), both published by Jossey-Bass, San Francisco, 1998.

2. Taking account of students' interests is an important element of many syntheses of effective designs for learning. See the focus on "learner-centered" designs in *How People Learn: Brain, Mind, Experience, and School*. National Research Council. Washington, D.C.: National Academy Press, 1999.

3. Gardner, H. *The Unschooled Mind: How Children Think and How Schools Should Teach.* New York: Basic Books, 1991.

4. Educational Technology Center. *Making Sense of the Future: A Position Paper on the Role of Technology in Science, Mathematics, and Computing Education.* Cambridge, Mass.: Educational Technology Center, Harvard Graduate School of Education, 1988.

5. Schwartz, J. "Symposium: Visions for the Use of Computers in Classroom Instruction." *Harvard Educational Review*, 1989, *59*(1), 51.

6. Many people doubt that such young children could be capable of the activities Kristi's students were. Yet students in her first- and second-grade classroom carried out this kind of work year after year. Children in her classes were a mixture of the kinds of children typical in the school, including some with diagnosed special needs; approximately half were eligible for Title I support. Throughout this case, and in the cases included in subsequent chapters, more information is provided about how Kristi developed a culture of inquiry and collaboration among her students, developed their abilities to write,

used "sound spellings" initially with young children and gradually taught them conventional spellings, and built their technological skills through engaging them in meaningful tasks with a wide range of tools. Kristi also gradually developed her own technical expertise, the range of tools available in her school, and her network of supportive relationships with colleagues, parents, and other volunteers. Teachers who doubt whether they or their own students could accomplish the kinds of learning that Kristi's classes have done might take heart from realizing that these stories build on many years of experimentation and refinement.

7. Eddy Spicer, D., and Huang, J. "Of Gurus and Godfathers: Learning Design in the Networked Age." *Education, Communication and Information, 1*(3), 2001, 325–358.

Understanding Goals and New Technologies

Most teachers have some sense of their goals as they begin a lesson, but few are entirely explicit about what they hope students will learn. Some goals may deal with particular academic content and skills, others with social behaviors, and still others with learning to use new technologies. Seldom do teachers and students fully understand their educational goals or see clear relationships among them, that is, see connections from specific skills developed in a particular lesson to the overarching goals and enduring value of a course, or see links from one subject matter to another, or from learning *about* technology to learning *with* technology.

Unless the teacher's goals are explicit, clear, and coherent, students are unlikely to achieve them. In this chapter, we advocate defining and publicizing clear goals that focus on important understandings and then using new technologies to improve the teaching and learning of these educational goals.

In the plethora and cacophony of required activities that teachers must orchestrate in the classroom, defining goals for students is complex. Teachers must consider their own overarching purposes and central passions as they define their core agenda. Most teachers are also expected to "cover" substantial curriculum content, which is often spelled out in standards or mandates. They may also be required to teach with specific textbooks or curriculum materials. And students may have to prepare for standardized tests or other

> *In this chapter, we advocate defining and publicizing clear goals that focus on important understandings and then using new technologies to improve the teaching and learning of these educational goals.*

Key Features of Understanding Goals

- Focus on big ideas
- Encompass multiple dimensions within or across subject matters
- Define coherent connections between any lesson or learning activity and broader overarching goals
- Are stated explicitly and shared publicly

---

accountability measures. All of these factors must be considered while defining understanding goals.

A further complication in defining educational goals is that pressure to implement new technologies may obscure the importance of using these tools to teach and learn important subject matter. Teachers may believe that simply engaging students with modern tools is worthwhile (to increase student motivation, develop students' technical expertise and twenty-first-century skills, or demonstrate up-to-date classroom practices), even if the technology is not helping students learn core curriculum topics. Teachers may incorporate technology into lessons in ways that provide some catchy entertainment with little or no contribution to learning.

New technologies are most beneficial, however, when they support and deepen students' understanding of important educational subject matter. In order to apply technology this way, teachers often need to reflect on their goals and become more explicit about exactly what they hope students will learn. Educational objectives, whether they deal with academic or technology issues, may reflect a teacher's deeper purposes, core values, and ultimate hopes for the students. But teachers rarely are encouraged to articulate these underlying goals, to link them directly with more specific learning objectives, or to share them publicly with students, parents, and administrators.

The Teaching for Understanding framework guides teachers to state clearly what they hope students will come to understand, both in the short term from a lesson

---

Factors That Must Be Coordinated When Defining Understanding Goals

- Teacher's deeper purposes, core values, and hopes for the students
- Mandates for curriculum content coverage
- Required textbooks and other curricular material
- Standardized tests and other accountability measures that require preparation time

---

or curriculum unit and long term over the period of a semester or year-long course of study. In Chapter One, *understanding* is described as a flexible capability to think with what one knows and to apply knowledge flexibly and appropriately in a range of situations. Goals aimed at this kind of understanding may encompass learning how to apply key concepts, methods of inquiry, how to appreciate purposes for learning, and forms of expression. Understanding goals do not focus solely on memorizing particular content objectives or practicing routine skills.

Suppose teachers want students to learn about technology. That's fine, so long as the goals aim for developing students' flexible understanding, not just isolated technical skills. For instance, teachers might want students to learn how to use a word processor or how to locate information on the Internet. But for what purpose? Well, perhaps so that students can produce better essays by being able to revise them more easily and so they can support their arguments with data and facts that they might not be able to find in the local library. Then the understanding goals around technology might be: "(1) Students will understand how to locate and critically evaluate information on the Internet that is relevant to their topic, and (2) students will understand how to develop a reasoned argument using evidence based on their own research. Phrasing the goals in this way reminds the teacher

TIP

Students need to master concrete objectives and basic skills, but these accomplishments are not an end in themselves. Learning specific skills and facts is a means of reaching larger ends.

Features of Effective Understanding Goals	**Features of Less Effective Goals**
• Focus on students' understanding of important knowledge, methods, purposes for learning, and forms of expression.	• Focus on isolated information or behaviors the teacher would like the student to learn.
• Address the essential aspects of the generative topic or discipline and help the student understand the substance as dynamic.	• Primarily address students' enjoyment rather than big ideas in the subject matter or discipline.
• Relate short-term goals for a lesson or unit to long-term overarching goals for a term or a whole course.	• Show no connection between the goals of a particular lesson and the larger purposes of a curriculum unit or entire course.
• May be phrased both as statements and as questions.	• Remain tacit or unclear to students.
• Invite students and colleagues into the process of generating goals.	• Are stagnant and set without discussion or collaboration with learners.
• Are revisited and revised for ongoing development.	

and the student that searching the Internet is not an end in itself but a means of identifying compelling evidence that supports or refutes one's argument. By focusing on such goals, students develop skills for using the Internet as they also learn how to find and evaluate relevant data. They develop technical fluency by using technology to accomplish meaningful work.

New technologies can be used in many ways to foster the teaching and learning of understanding goals. At the same time, clear understanding goals help teachers, students, and other key participants in the educational process (such as parents, technology specialists, school administrators, and policymakers) ensure that technology provides significant educational leverage. The following case study illustrates ways that clear understanding goals and new technologies can be mutually reinforcing.

Case Study: The Water Habitat Project

For over a decade, in all seasons of the school year, the six-, seven-, and eight-year-old students in Kristi's primary classroom made treks to a nearby city park pond for science studies. Each field trip yielded new data for longitudinal research on the pond. In the early years, the project looked like many elementary school field activities in science. Students observed aquatic plants and animals, visually gauged water quality in the pond, and recorded the water pH and temperature. They wrote their field trip observation data in spiral notebooks and made photo journals using 35mm pictures of each field trip. By comparing findings from their trip with the previous year's written observations and photo journals, students noted changes and trends over time.

Kristi's overarching goal in the Water Habitat Project was for students to understand environmental science by being scientists. She wanted students to understand the strategies, methods, techniques, and procedures that environmental scientists use in today's world.

Kristi's overarching goal in the Water Habitat Project was for students to understand environmental science by being scientists. She wanted students to understand the strategies, methods, techniques, and procedures that environmental scientists use in today's world.

More specifically, students would understand how to collect, record, and analyze data to construct scientific understanding. As a primary teacher, Kristi also wanted the environmental science work to provide authentic, real-world content for these young students to understand reading, writing, and communicating.

How New Technologies Transformed the Project

During the 1993–94 academic year, as students were studying the longitudinal pond data, they realized dramatic changes were taking place in this habitat. The island in the pond—a nesting habitat for ducks and geese—was disintegrating. The water was increasingly being polluted with debris from nature and people. The students' inquiries about the habitat escalated as they wanted to know what was causing the changes. Kristi realized that students' intensifying study of the pond provided an excellent opportunity for them to learn about both science and literacy by documenting and communicating their observations, questions, and concerns about this habitat.

Kristi realized that students' intensifying study of the pond provided an excellent opportunity for them to learn about both science and literacy by documenting and communicating their observations, questions, and concerns about this habitat.

During this same school year, Kristi's classroom acquired a computer and printer for word processing. With these tools, students could record pond observations faster and immediately print multiple copies of their documents for the class to read and discuss. The class became excited about the details they were recording about the pond. They also met with the City Parks Department supervisor to ask questions about the pond changes. They were becoming collaborative environmental scientists through scientific inquiry about their local water habitat.

In January 1994, Kristi added a modem and Internet connection to the classroom and joined the K–12 International Education and Resource Network (iEARN). This organization links teachers and students in schools around the world to help them learn from one another through on-line collaborative curricular projects and collaborate on making positive contributions in their communities using what they learn together.

Initially, Kristi's primary students used e-mail to participate in the iEARN global water habitat study. They exchanged pond reports as e-mail messages with other schools around the United States, Costa Rica, Russia, Australia, and Argentina.

They exchanged pond reports as e-mail messages with other schools around the United States, Costa Rica, Russia, Australia, and Argentina.

The e-mail messages prompted students to ask questions about one another's habitats. Comparisons of data enabled students to identify and understand common and diverse aspects of water habitats around the world.

As Kristi realized what her students were learning through e-mail communication with distant peers, her understanding goals began to expand beyond what had previously seemed possible. Her students were coming to understand

- How environments are interconnected around the world
- How scientists collaborate to build scientific knowledge by sharing their data and critical thinking through on-line inquiry
- How global peers cared about their own and one another's water habitats

Kristi also noticed that students invested more effort in writing clearly when they were creating e-mail messages for their peers; they appreciated that they had to explain their ideas carefully to avoid being misunderstood.

In spring of 1994, another new technology further transformed the water habitat project: the class used CUseeMe videoconferencing software to present their

Kristi also noticed that students invested more effort in writing clearly when they were creating e-mail messages for their peers; they appreciated that they had to explain their ideas carefully to avoid being misunderstood.

pond habitat project to other K–12 students who were attending an environmental conference at the American Museum of Natural History in New York City. When Kristi's students shared their habitat data and their concerns about the deteriorating conditions of the pond, a New York City youth suggested, "You need to get a lot more people to care!"

That on-line comment from a fellow student spurred the class to meet again with local officials and ask what could be done for the pond. The City Parks and Recreation Commission invited the class to make a presentation about their concerns. Using longitudinal scientific data and photo journals, their writing, artwork, e-mail communication, and excerpts from the videoconference conversation, the children explained the situation at the pond and requested that the city consider restoring the island and the water quality.

Influenced by the strength and integrity of the students' scientific understandings, as well as their presentation skills, the Parks and Recreation Commission voted to allocate funds for the pond restoration. From this experience, students learned the value of developing and presenting scientific findings and the possibility of applying their understanding of science to civic action. In subsequent years, as the restoration proceeded, Kristi's students continued to work with the city officials to monitor the pond habitat, thus extending the longitudinal research conducted by prior classes.

The use of new technologies in the pond habitat project continued to evolve. Students learned to use digital images, Web sites, video production, and the Washington State K–20 videoconferencing network to document and disseminate their pond habitat findings to multiple audiences. Students took digital photographs during their field trips, which they could view immediately when they returned to the classroom. These images prompted discussions and analyses, as students wrote notes about their pond observations. Sometimes the students made additional significant observations while viewing the digital photos that they had missed during their visit to the site. Students used digital images with text in their pond journals to provide rich documentation of the water habitat. They used printed copies of their pond journals to practice reading about science, both in the classroom and at home.

The class synthesized material from these journals to develop a Web site of longitudinal water habitat data, photos, and reports. The site became a science resource for Kristi's classes, as well as a means of sharing their science work with the Parks and Recreation staff, their families, other members of the community via computers at home or in the public library, and their colleagues around the world.

The Water Habitat Project curriculum coupled traditional field-site science methods for observing a dynamic living ecosystem with innovative uses of new technologies to transform education from learning about science to being scientists. Kristi's young students learned in their first years of school how to do science as a process of using multiple technologies to gather systematic data, collaborate to share and analyze data with other members of a scientific community, and develop and disseminate scientific findings in compelling ways that promoted beneficial civic action. With a clear focus on understanding goals, Kristi addressed multiple required curriculum standards while engaging her students in authentic work.

To learn more about the process and results of this work, visit the class Web site at http://learnweb.harvard.edu/ent/gallery/pop3/pop3_1.cfm.

Key Features of Understanding Goals

This case illustrates several key features of understanding goals. First, understanding goals are central to the subject matter and *focus on big ideas,* beyond memorizing facts or rehearsing routine skills. Such goals may encompass *multiple dimensions* within or across subject matters. After reviewing many different frameworks for categorizing educational goals, members of the Teaching for Understanding project[1,2,3] defined four strands or dimensions of goals. These dimensions are as follows:

- *Knowledge,* including key theories and concepts, such as understanding a biological habitat as a dynamic system

- Discipline-based *methods* for building and validating knowledge, such as using scientific inquiry and analysis to gather and interpret data

- Appreciating the *purposes* for learning, including developing autonomy to use what one knows, such as understanding why accurate science methods and data are essential for building knowledge and how to apply scientific inquiry in advocating for responsible environmental actions in local to global communities; and

- *Forms of expression,* such as using visual arts (including digital photos and video), numerical charts, and written and spoken language, to explain ideas in clear and persuasive ways.

The key points are that the understanding goals (1) focus on important, rather than trivial, aspects of the subject, (2) target performance-based understanding, including a flexible capacity to think and apply knowledge, not simply learn facts, definitions, formulas, or routine skills.

Memorizing the steps in some version of "the scientific method" is not an understanding goal. In contrast, Kristi's goals were for her students to understand how and why to practice scientific methods of inquiry. Understanding goals require

With a clear focus on understanding goals, Kristi addressed multiple required curriculum standards while engaging her students in authentic work.

TIP

Understanding goals focus on important, rather than trivial, aspects of the subject and target performance-based understanding, including a flexible capacity to think and apply knowledge, not simply learn facts, definitions, formulas, or routine skills.

going beyond the information given to make connections, create a solution, or apply knowledge in a new context.

Another important feature of understanding goals is that they *define coherent connections* between any lesson or learning activity and broader overarching goals. In this way, both teachers and students can easily track the reason for any particular exercise in relation to the larger purposes for studying and learning. In the Pond Habitat project, Kristi's overarching goal was for students to understand how to do environmental science by being scientists. This large goal encompassed multiple sets of understanding goals in science, as well as in mathematics, social studies, and literacy and communication. Within science, for example, Kristi's goals for her students included understanding how the local pond habitat was a dynamic system, how to make and record accurate observations and measurements, how to present scientific findings in clear and compelling formats, and why scientific methods are necessary for building reliable knowledge and for making persuasive arguments. Students practiced skills such as correcting spelling errors and improving their keyboarding speed by using these skills for meaningful purposes.

One final important point to make about understanding goals is that teachers *state them explicitly and share them publicly.* In order to articulate their goals, most teachers have to invest considerable thought and effort in reflecting on their subject matter, their students, and the larger purposes they hope to accomplish with their students. Teachers often need to surface tacit knowledge buried in the back of their minds. This effort pays off as teachers share their goals with learners, parents, and other stakeholders. Instead of trying to guess what the teacher wants, students can invest their effort in marshalling their resources to reach a stated destination.

Part of Kristi's regular practice is to state her goals clearly for every lesson and to clarify these understanding goals with her students in response to their comments, questions, struggles, and breakthroughs in learning progress. Kristi also regularly invites students to reflect and comment on their progress toward goals because one of her overarching goals is that students will understand how to monitor and promote their own learning.

Using Technology to Enhance Understanding Goals

Clear understanding goals can serve as a guide for using technology wisely. At the same time, new technologies can provide significant advantages in making understanding goals vivid, memorable, and accessible to learners. Being clear about how new technologies directly support progress on understanding goals helps teachers ensure that technology provides significant educational leverage.

One of Kristi's goals was that her students would understand how to participate in a scientific community that included former students whose data would provide longitudinal evidence about change, peers in other parts of the world who were gathering comparable data in their own settings, and experts who could provide specialized information about the students' investigations. New technologies provided several benefits in achieving this ambitious goal.

Kristi's students used digital photographs and videos to capture images of the pond habitat. They compared their images with those made in previous years, in addition to comparing text and quantitative data, to reveal longitudinal changes in the habitat. By capturing their data in multiple, comparable formats, students could progressively build a longitudinal database and contribute to a growing body of scientific research.

The students in Kristi's class used word processors and digital images to create science journals. By writing their understandings of this water habitat individually and in small groups with a word processor, the students could easily combine products into a whole class report, add digital images, and collaborate on editing the document until it clearly and accurately expressed their ideas. Because these materials were captured in digital forms, the products could easily be posted to e-mail and incorporated into the class Web site. Networked technologies such as e-mail and the Web allowed Kristi's students to collaborate with peers around the world. Exchanging data and ideas with global classmates allowed Kristi's students to understand the importance of accurate methods and clear communication and to appreciate the nature of science as a collaborative process of inquiry. The comparison of

How New Technologies Can Support Understanding Goals

Multimedia Technologies	Enable students to express their understanding through multiple formats, including audio, video, images, and text
Graphing Calculators	Allow students to understand the relationship between a symbolic expression and the graph of a function
Scientific Simulations	Enable learners to grasp difficult concepts by "showing" events that are not actually visible
Digital Facsimiles	Provide access to rare historical artifacts; help students appreciate the importance of primary sources and the nature of bias in historical interpretation

Identifying understanding goals helps teachers guide students' use of technology to focus on key priorities. New technologies can provide valuable educational leverage on understanding goals that would be difficult or impossible to achieve otherwise. In this way, understanding goals and new technologies can be synergistic educational innovations.

data with distant colleagues also helped students understand variations and similarities across different pond habitats. Multimedia technologies, including digital video, the class Web site, and videoconferencing, allowed students to present their scientific findings in compelling ways that led to important civic actions both in their own community and in distant places.

Many other kinds of technologies can also help make understanding goals more clear and accessible to learners. For example, graphing calculators display mathematical functions in multiple forms so that students can readily see the relationship between the symbolic expression and the graph of a function. Numerous simulations of scientific phenomena, such as heat transfer at the molecular level, enable learners to grasp difficult concepts by "seeing" depictions of events that are not actually visible in ways that highlight important concepts. Access to digital facsimiles of rare historical artifacts can help students appreciate the importance of primary sources and the nature of bias in historical interpretation. Web sites like the one maintained by the Folger Shakespeare Library (www.folger.edu) bring together a wealth of resources for teaching literature, such as multiple versions of Shakespeare's works and video clips of different actors' interpretations of the same lines in a play. With such resources, students can appreciate literary works and theatrical interpretations as products and promoters of human invention.

Identifying specific understanding goals helps teachers guide students' use of technology to focus on key priorities rather than drift into using the technology as an entertaining but frivolous or distracting activity. At the same time, new technologies can provide valuable educational leverage on understanding goals that would be difficult or impossible to achieve otherwise. In this way, understanding goals and new technologies can be synergistic educational innovations.

For Kristi, clarifying the goals she hoped students would come to understand was a gradual process.

Planning and Teaching Understanding Goals with New Technologies

The Water Habitat project may sound wonderful, and the key features of understanding goals may be clear, but how do teachers actually define understanding goals and use them to focus their work with new technologies? It is a process that requires thinking about what students need to understand and how new technologies might be employed to support their learning. For Kristi, clarifying the goals she hoped students would come to understand was a gradual process.

She reflected on her own heartfelt purposes, studied the curriculum standards and required tests that she was expected to address, and considered the important

The WebQuest Web Site at San Diego State University: Learning About Technology Tools to Advance Students' Understanding[1]

The Internet offers vast resources to enrich learning activities, but teachers are challenged to help students use these resources in ways that develop and demonstrate understanding of target goals. The WebQuest Web site at San Diego State University (SDSU) supports teachers and students in making effective use of the Internet.

A WebQuest is a Web site that organizes students' work and use of on-line resources for a curriculum unit. The WebQuest template includes an introduction to hook the student, a task with clear goals, a process that describes what students will do with hyperlinks and includes resources and tools for organizing their work, evaluation guidelines with criteria to meet the goals, a conclusion that encourages students to reflect on their work, and a teacher page with information that helps other teachers implement the WebQuest. With these elements, WebQuests keep throughlines and understanding goals visible, sequence challenging tasks and performances, make expectations for ongoing assessment clear and public, and provide hyperlinks to carefully selected on-line resources so that students stay focused on understanding goals.

The WebQuest site includes many examples of WebQuests that teachers have developed with this template. Teachers may use an existing WebQuest, modify it for their own purposes, or create an entirely new WebQuest. Students themselves can create WebQuests. Because a WebQuest is published on-line, students can work on it wherever they have access to the Internet.

The SDSU WebQuest site offers a number of tools to help educators learn to use WebQuests and other technologies to teach for understanding. The site provides training materials, including theoretical and practical articles, tutorials, links to on-line workshops, and a matrix of classroom-tested and rated examples of WebQuests organized by subject and grade level. Visitors to the WebQuest site can connect with other teachers in an interactive forum, join threaded discussions to pose questions, share ideas and research, comment on WebQuest designs, find resources (such as non-English WebQuests), and work through practical issues. The WebQuest site is a rich resource to build teachers' understanding of how to use the Internet to advance students' learning.

Resources

The WebQuest Page at San Diego State University
http://webquest.sdsu.edu/

International Society for Technology in Education (ISTE)
Standards for Teachers
http://cnets.iste.org/teachers/t_stands.html

Ozline: Helping Educators Work the Web
http://www.ozline.com/

Filamentality
http://www.kn.pacbell.com/wired/fil/

[1]This vignette describes the work of Bernie Dodge.

learning that her students were achieving. Analyzing these sources and discussing her goals with colleagues, including scientists and other subject matter specialists, helped her define and progressively refine clear goals.

Uncovering Core Purposes and Overarching Goals

At root, understanding goals are embedded in the overarching values that motivate teachers to become educators.

For Kristi, helping her students learn how to be responsible citizens in their community and in the larger world is key. Learning how to read with understanding, to write in ways that convey ideas accurately and persuasively, and to carry out systematic scientific inquiry are important means toward the larger goal of contributing carefully and creatively to the global community. For those who teach teachers and other educators, overarching goals may include developing educators' capacity and the dispositions to continue participating in a reflective, collaborative professional community.

Whatever particular knowledge and skills teachers may want their students to develop, these specific goals are stepping-stones to larger purposes. In beginning to clarify understanding goals, it is worthwhile to take time to contemplate one's own core values and the overarching goals one hopes students will achieve. These long-term goals, which may endure over a semester or a year, provide a focus for the more specific plans for curriculum units or lessons. Within the Teaching for Understanding project, these overarching understanding goals were sometimes called throughlines because they provided steady direction and focus through an entire course.

Defining Understanding Goals for Units and Lessons

Clarifying understanding goals for a particular curriculum unit or project requires making connections between overarching goals (or throughlines) and more specific—often mandated—curriculum goals. When Kristi began to think about her goals for the Water Habitat project, she asked first: What is the full range of possibilities for learning that this project can offer? She realized that her students could learn about science, language arts, mathematics, social studies, art, and new technologies through this project. In each of these subject matters, Kristi defined major goals that connected both to her core purposes and to the subject matter goals that she was expected to address. For example, one of her unit-level goals was, "Students will understand how to use science, math, literacy, arts, and telecommunications skills and knowledge to make positive contributions in local to global environments."

Goals at this level serve both to anchor educational plans in the teachers' overarching priorities and to help students appreciate the larger purposes of their studies. But these large goals must then be specified in more detail in order to serve as guides for planning particular lessons.

TIP

One way to surface such goals is to imagine a student returning to visit his teacher five years after leaving her class and saying, "What I'll always remember about learning with you was. . . ." How might the sentence end? What kinds of learning do we most hope students will carry away with them? What are the most important understandings that students need in order to succeed in the world?

A next step is to review the specific learning goals or curriculum standards at district, state, and national levels for relevant subject matters and grades. Both Kristi's school district and the state defined required curriculum goals. She also reviewed national curriculum standards for elementary-grade students.

Translating these materials into understanding goals is not a simple matter, however. One problem is the "thud factor," as one teacher called it—that is, so many standards that teachers cannot possibly address all of them. Another problem is that many standards focus on memorizing particular facts, learning about a specified method, or practicing skills rather than on requiring students to think with what they know and stretch their minds. Similarly, as teachers consider the required tests that will be used to judge their students' achievement, they may realize that most of the items test students' recall of isolated facts and formulas rather than real understanding. How should teachers respond to such a welter of standards and test items? How might they connect or combine disparate standards into true understanding goals?

In reviewing required curriculum guidelines while keeping overarching purposes and the criteria for understanding goals in mind, teachers may begin to see clusters of goals around the different dimensions of understanding described earlier in this chapter: knowledge, methods, purposes, and forms of expression. For example, Kristi realized that there were several different scientific *knowledge* goals that she could consolidate, as follows:

1. Students will understand how water habitats, such as the local pond habitat they study, are dynamic systems of changes and how these changes happen over time because of climate, topography, geography, and interactions and interdependencies of species populations, including human effects and human management actions. Similarly, she consolidated a number of curriculum standards from multiple subject matters into a more specific goal that addressed *methods* of inquiry and analysis, as follows:

2. Students will understand how to use the math skills of computation, graphing, and creating tables along with literacy-communication tools (including digital images, video production, e-mail, Web sites, and videoconferencing) to accurately document, analyze, and collaborate with local and global peers on understanding water habitat observations and data.

Kristi also wanted her students to understand the larger *purposes* of their schoolwork in relation to the real world and to develop expertise in using several new kinds of information and communication tools. She consolidated these aims into the follow understanding goal:

3. Students will understand how to use on-line communication technologies, including e-mail, Web sites, and videoconferencing, as well as multimedia tools, to present the results of their scientific inquiry in ways that initiate and implement community action and service efforts to maintain and restore water habitats.

TIP

Clarifying understanding goals for a particular curriculum unit or project requires making connections between overarching goals (or throughlines) and more specific—often mandated—curriculum goals.

In addition, Kristi clustered other curriculum standards and goals around technological literacy into the formulation of a goal that focused on understanding *forms of expression* both for individual students and for groups, as follows:

4. Students will understand how to use the new technologies of e-mail, Web sites, digital images, and videoconferencing communication tools to collaborate with global peers in analyzing longitudinal water habitat observations or data and explaining implications of changes in pond habitat, including comparison of commonalities and diversities of changes in water habitats studied by other schools.

These four dimensions of understanding—*knowledge, methods, purposes,* and *forms*—provide useful categories for dealing with the thud factor by consolidating or clustering myriad small goals into important understanding goals. Reviewing the criteria for understanding goals may help teachers combine and reword rather low-level objectives into more thought-provoking understanding goals. Sometimes simply changing a few words signals a big shift in aims. For example, "learning the steps in the scientific method" points toward memorizing some specific sequence, whereas "understanding how to apply scientific methods to gather, analyze, and present knowledge" suggests that learners will exercise judgment and appreciate a range of aspects of scientific modes of inquiry. Teaching students specific concepts and skills can then be accomplished in the context of developing their higher-level understanding.

Revising Understanding Goals

Weaving together one's own priorities with curriculum requirements into a clear progression, or nested set, of understanding goals is a challenging process. Most teachers find that they need to talk with other thoughtful educators, and perhaps with subject matter specialists, to carry out this work.

Once teachers have spelled out a version of their understanding goals, they can begin to use it as a guide for designing learning activities and assessments, as well as for planning effective uses of technology. But the process of defining goals will continue as teachers discover the potential of new technologies and observe what their students can achieve. Kristi continued to review, clarify, and refine her goals as she taught the Water Habitat unit year after year.

The revisions were prompted in part by students who developed understandings that surpassed even her ambitious expectations. As they used e-mail, the Internet, digital video, and other tools, students demonstrated capacities to conduct scientific inquiry, analyze their findings, and develop compelling presentations for local and global audiences that far exceeded the curriculum standards for primary students. So Kristi looked to the standards for older grades to help her articulate more advanced goals to match her students' capacities.

TIP

Talking with colleagues can help bring into clear focus ideas that are partially buried in the back of one's mind. Conversations with specialists in relevant fields of study can help teachers formulate key concepts and discipline-based modes of inquiry in clear and specific terms.

Integrating New Technologies with Understanding Goals

Integrating new technologies to improve the teaching and learning of understanding goals is also a cyclical process of planning, trying out a new approach, reflecting on how it worked, and revising. Although defining goals and using new technologies to achieve those goals are intertwined activities, they are described separately here to highlight important aspects of both processes.

Kristi continued to review, clarify, and refine her goals as she taught the Water Habitat unit year after year.

There are at least four issues to address when integrating new educational technologies into one's practice: (1) selecting appropriate technologies, (2) planning effective ways to gain significant educational advantage from the technology, (3) preparing oneself and the students to use these new tools, and (4) organizing both access to the tools and educationally effective interactions of learners with the tools and with other members of the learning community.

Selecting Appropriate Technologies

In the Water Habitat Project, one of Kristi's primary goals was to engage students in experiencing science as a collaborative process conducted through dialogue with fellow scientists and through presentation of findings to appropriate audiences. Networked communication technologies, like e-mail and the Web, seemed obvious tools for enhancing this kind of scientific dialogue. E-mail offered a fast means of exchanging data, questions, and findings with fellow learners. Creating a Web site helped the class consolidate various kinds of findings in vivid ways and make this material easily accessible to several audiences. Word processors helped students collaboratively draft, revise, and produce written documents. As Kristi and her students began to work with these computer-based technologies, they discovered additional, related tools that were useful to them, including video production and video conferencing.

Teachers may learn about technologies from colleagues or from seeing how other professionals use them. Sometimes a technology specialist or administrator informs

---

Integrating New Technologies into Teaching Practice: Issues to Address

- Selecting appropriate technologies
- Planning to gain significant educational benefits
- Preparing teachers and students to use these new tools
- Organizing access to the tools and orchestrating effective interactions

Archetypal Heroes in Western Civilization[1]

Teachers find that students have difficulty differentiating celebrity from heroism. In this humanities unit, high school students use the Internet and visual organizing software to analyze hero archetypes. The unit goals are to understand (1) the purpose and function of heroes in myth, history, and culture, (2) how heroes embody the aspirations, fears, conflicts, and triumphs of a community and a civilization, and (3) how heroism differs from celebrity.

In a series of introductory lessons related to these overarching goals, students read about archetypal heroes and explore representations of heroism from history, philosophy, criticism, drama, literature, art, current events, and pop culture, using a variety of Internet resources. Because the Internet itself can be a medium of myth-making, students refer to guiding questions from their teacher in order to refine their skills for assigning information as they write essays exploring the purpose and function of heroes in cultural and historical context.

The class then works together to negotiate a definition of the universal qualities of heroism. As they collaborate, students use a flexible visual-organizing software tool, such as Inspiration®, to keep unit goals in mind and to capture and map their evolving ideas and understanding of heroes. Students generate and rearrange their thoughts and arguments about celebrities and heroes, using Inspiration to organize ideas through words, images, and animations with hyperlinks to digital source materials (text, images, video or sound files) that provide evidence and illustrate how a figure meets the class's criteria for heroism. The software supports understanding by allowing students to save, compare, and assess iterations of their work throughout the progress of the unit.

Each student then chooses and researches one heroic figure in greater depth, preparing a reasoned argument for a class debate. What function does the figure play in a social and historical context? Which literary archetype does the figure represent? Does the figure meet or challenge the class's definition of the universal qualities of heroism? Is the figure a hero or a celebrity? Students use Inspiration to create and present visual maps that summarize their rationales.

As a culminating performance to demonstrate understanding of heroism, each student invents his or her own heroic character, myth, and cultural context. Students may present their invented heroes in dramatic or dance performance, visual artwork, sculpture, essay, oral presentation, epic poem, computer game, or comic strip. Presentations are captured with multimedia technologies so that teacher and students may reflect on and assess their understanding of heroism and celebrity from history to present times.

Resources

Heroes Ancient, Modern and Mythic: A Bibliography
(Published by the Barron Prize for Young Heroes)
http://www.barronprize.org/tabarron/heroes.html

Center for Story and Symbol
http://www.folkstory.com/

Heroism in Action (A ThinkQuest Entry)
http://library.thinkquest.org/C001515/welcome/

Time Heroes and Icons
http://www.time.com/time/time100/heroes/

My Hero
http://myhero.com/home.asp

The Hero's Journey: An environment to explore the classic mythical story structure and to create your own stories.
http://www.mcli.dist.maricopa.edu/smc/journey/

iEARN Project: My Hero Project
http://www.iearn.org/projects/myhero.html

---

[1]This vignette is based on the work of Monica Hiller.

teachers that a new tool is available and that teachers should integrate it with their practice. Many Web sites consolidate information about educational technologies useful for different subject matters and ages of learners.

Planning to Gain Significant Educational Benefits

A clearly articulated set of understanding goals provides an explicit basis for deciding whether a new technology is appropriate and how it might offer significant educational benefits. Kristi realized that the International Education and Resource Network (iEARN) provided a wealth of opportunities to connect her classroom with the wider world. Her specific goals about science, communication skills, and collaboration to serve local and global communities helped her to decide how exactly to take advantage of iEARN's on-line resources.

Reviewing the dimensions and criteria for understanding goals may help to pinpoint particular educational advantages of a new technology. With these considerations in mind, teachers may see ways to use new technologies that will improve students' understanding of key *concepts, methods* of inquiry, *purposes* for learning, and *forms* of expression. How might technology help students distinguish between important and closely related concepts, such as species interdependence and species interaction? Can it provide students with a means to see important connections, such as patterns in longitudinal data or common elements of multiple sites of water habitats? Can it help students appreciate a fundamental yet difficult concept such as different points of view between the writer and the reader of documents? The main point is that teachers should have one or more explicit reasons that are linked to stated understanding goals for integrating any technology into their lesson plans.

Of course, the potential of new technologies is often quite varied and difficult to anticipate. Discovering unexpected benefits after starting to use innovative tools is likely, and teachers can always revise their goals. But it's still important to have clear, educational goals stated initially so that public goals can serve to guide students toward important understandings. Otherwise, teachers and students are likely to waste too much time or use the tool for decorative touches rather than valuable learning.

Preparing Teachers and Students to Use These New Tools

Certainly, teachers need to learn something about a new tool before trying to teach with it or asking students to use it. But many teachers believe they have to invest much more effort in learning the technology than in fact is necessary. Teachers don't have to work their way through the entire manual or user tutorial in advance. If

> **TIP**
>
> As teachers review technology options, they need to ask, "What tools are feasible for me, for my students, in our context?" and "Which tools are really valuable for my goals?"

they understand the basics of using a word processor, for example, or know how to make a simple diagram with a piece of software like PowerPoint® or Inspiration®, that may be enough to get started.

Teachers may need to allow time to show students how to work effectively and responsibly with new technology. The first time students are introduced to a new tool, this investment in technology training may seem too time consuming. If a teacher expects to use a complicated tool or software only once, then it may be worth reconsidering whether this application of technology is really beneficial enough to warrant the time it takes to teach. If the same tool may be used for other lessons, however, this initial preparation will pay off many times.

Students themselves can support their own learning about new tools. They are often able to learn about technology faster than older people, and usually some students want to serve as technical assistants for their peers. So long as the experts are not always the same students, and the experts are also helped to advance their own understanding, this arrangement of peer assistants is good for everyone.

Kristi also invited other assistants to help when students were working with complex technologies and undertaking demanding work in small groups. Realizing that the ratio of 1 adult to 20–25 (or more!) young children cannot adequately provide the support needed for students to achieve understanding goals, she invited parents, student teachers, and science specialists to help. They assisted on field trips and in the classroom by listening to and mentoring students' discussions and by recording students' developing understandings. Technology-savvy parents also provided valuable assistance when students and the teacher were learning new technologies. To teach for understanding with new technologies in her classroom, Kristi never hesitated to ask for help or to ask lots of questions. Using new technologies is not a matter of learning something once but rather a process of

---

Preparing to Use New Technologies

- Many teachers believe they have to invest much more effort in learning the technology than in fact is necessary. Teachers don't have to work their way through the entire manual or user tutorial in advance.

- If a teacher expects to use a complicated tool or software package only once, then it may be worth reconsidering whether this application of technology is really beneficial enough to warrant the time it takes to teach.

- Using new technologies is not a matter of learning something once but rather a process of continual learning while discovering new software features that are relevant to one's educational goals.

continual learning while discovering new software features that are relevant to one's educational goals.

Organizing Access to the Tools and Orchestrating Effective Interactions

There is no one correct or best way to use any technology. Deciding how many computers or other devices are needed for a class depends partly on the teacher's own preferences within the constraints and resources of a particular situation. Some teachers prefer to use a single machine, projected or connected to a large screen, as an interactive chalkboard with their whole class. Many teachers find that students work more effectively in pairs or trios than when they work alone. Students can be beneficial resources for one another in exchanging ideas about challenging work, providing peer feedback, and keeping one another on task. Often the rhythm of students' work with technology is most effective if they intersperse time on-line with time offline, that is, time to prepare, review their work individually, or consult with the teacher or other classmates. Alternative educational designs suggest different patterns for using technology and for grouping students.

Of course, decisions about how to orchestrate students' interactions with technology and with other people depend partly on the way resources are deployed and scheduled within one's own setting. Some schools corral their computers into special laboratory settings. These labs can be efficient for lessons in which all the students need to work on computers at one time. Computer labs may also be staffed with technical assistants, which is helpful especially if the assistants understand the teacher's curricular goals. Sometimes, schools choose the lab arrangement because it provides an easy way to equalize access to computers (for example, every class might get forty-five minutes per week), even though this arrangement may not be the most educationally effective structure. If teachers are assigned a time in the lab that does not serve their educational goals, they should consider negotiating an alternative schedule with colleagues.

Some schools supply computers to individual classrooms. Frequently, the classroom has only a single computer. Even one machine can be a valuable teaching aid if it is connected to a large screen or projector. Some schools rotate clusters of portable computers into classrooms as teachers request them for particular projects. As students' demonstrate academic achievements with new technologies, their accomplishments can become a basis for teachers to acquire additional computers through grants, educational awards, and generous donors.

Any of these varied technology arrangements can be used to enhance students' understanding, but they all require careful planning about goals and about the use of technology. As teachers plan the interaction of their students with technology, they must consider realistically the sequence of work that they must

---

Organizing Effective Uses of Technologies

- Often the rhythm of students' work with technology is most effective if they intersperse time on-line with off-line time to prepare, review their work individually, or consult with the teacher or other classmates. Alternative educational designs suggest different patterns for using technology and for grouping students.

- There is no one correct or best way to use any technology. Deciding how many computers or other devices are needed for a class depends partly on the teacher's own preferences within the constraints and resources of a particular situation.

- As teachers plan the interaction of their students with technology, they must consider realistically the sequence of work that they must accomplish and the realities of their own situation. It is better to start with a relatively simple plan and then make it more complex after "ironing out the wrinkles" in the first trial.

---

accomplish and the realities of their own situation. It is better to start with a relatively simple plan and then make it more complex after "ironing out the wrinkles" in the first trial.

When the teacher and students are clear about their goals and about how new technologies can support achievement of those goals, they are more able to focus their efforts. Making goals explicit and public ensures that other important actors—parents, technology specialists, and administrators—play their part in helping students develop important understandings.

Questions for Reflection

1. What large, overarching purposes do you hope to achieve with your students?

2. Within the subject matter that you teach, what specific goals do you want students to understand through their study of the generative topic you have chosen? Consider goals related to knowledge, methods of inquiry, purposes for learning, and forms of expression.

3. What new technologies that are realistic for you to try might help students achieve these goals?

NOTES

1. Boix-Mansilla, V., and Gardner, H. "What Are the Qualities of Understanding?" In *Teaching for Understanding: Linking Research with Practice*. Wiske, M. S. (ed.), San Francisco: Jossey-Bass, 1998.

2. See work done in the Knowledge Forum (http://www.knowledgeforum.com/K-12/products.htm), which uses new technologies to support students in building knowledge together with peers, teachers, and other colleagues, including subject matter specialists. See also "Engaging Students in a Knowledge Society" (Scardamalia, M., and Bereiter, C. *Educational Leadership,* 1996, *54*(3), 6–10).

3. For more information about how teachers clarify understanding goals, see the following: "How Do Teachers Learn to Teach for Understanding?" (Wiske, M. S., Hammerness, K., and Wilson, D.) and "How Does Teaching for Understanding Look in Practice?" (Ritchhart, R., Wiske, M. S., with Buchovecky, E. and Hetland, L.). Both chapters are in *Teaching for Understanding: Linking Research with Practice*. Wiske, M. S. (ed.), San Francisco: Jossey-Bass, 1998. For more about this process, along with specific examples of understanding goals, see *The Teaching for Understanding Guide* (Blythe, T. and Associates, San Francisco: Jossey-Bass, 1998). Two Web sites that provide extensive resources about Teaching for Understanding, as well as an on-line tool to support the design of curriculum with this tool, are "Active Learning Practices for Schools" (ALPS) at http://learnweb.harvard.edu/alps and "Education with New Technologies" (ENT) at http://learnweb.harvard.edu/ent.

Performances of Understanding and New Technologies

The Teaching for Understanding project defined *understanding* as a capability to think and perform flexibly with your knowledge, for example, to solve a problem, present ideas in clear and compelling ways, apply concepts by using them to describe or explain something like a poem, an historical event, an organ system, or an interaction with a classmate.

The project called such activities *performances of understanding* and found that they were effective means of *developing* as well as *demonstrating* understanding. If learners are going to think for themselves and become able to apply what they know in appropriate and creative ways, then the learning process must engage learners in just this kind of active thinking. Of course, learners also need to take in new information, perhaps by listening to a teacher's presentation or by reading. But effective teachers make sure that students spend a large proportion of the time involved in actively using and stretching their minds, not just passively receiving knowledge that others have created.

Not all forms of discovery learning or "project-based" curriculum actually involve true performances of understanding. Students may be involved in activities that don't really require them to think or to learn about priority goals. Projects may be "hands-on" without requiring much "minds-on" work. For example, using a word processor to improve the appearance of a finished written product may interest students and keep them busy without advancing their understanding of the target goals. Indeed, many complex projects

The Teaching for Understanding project defined understanding *as a capability to think and perform flexibly with your knowledge.*

Key Features of Performances of Understanding

- They develop and demonstrate understanding of target goals.
- They *require students to stretch their minds*—to think beyond what they have been told, confront their usual ideas and attitudes with a more critical perspective, and combine or contrast ideas in ways they have not done before.
- They build up understanding through a *sequence of activities* that gradually transfer autonomy and responsibility to learners.

---

require a lot of time for tasks that are routine or quite peripheral to the main curricular agenda. Thus one prime feature of performances of understanding is that they develop and demonstrate *understanding of target goals.*

Another feature is that performances of understanding *require students to stretch their minds*—to think beyond what they have been told, confront their usual ideas and attitudes with a more critical perspective, and combine or contrast ideas in ways they have not done before.

In order to engage and stretch all learners, a good sequence of performances begins with *introductory performances* that connect to students' interests and initial levels of understanding. Then the learning activities build up students' understanding through a series of *guided inquiry performances* that initially include more guidance from the teacher and gradually transfer more autonomy and responsibility to the learners—all the time requiring students to "go beyond the information given"[1] and use their minds. Students often draft and revise work during this process, as they gradually come to understand the qualities of good work and develop the capacity to produce work that meets these criteria. During a sequence of guided performances, teachers may tailor the amount of direction and support they provide to individual students, depending on each student's initial level of understanding and particular needs. A curriculum unit may end in a *culminating performance* that requires learners to synthesize a range of understandings, perhaps encompassing different subject matters and developed over a considerable period of time but still focused on the target goals. Effective teachers design performances that allow their learners to use what Howard Gardner calls "multiple intelligences,"[2] that is, various ways of thinking and forms of expression that may include verbal, mathematical, visual, musical, movement, introspective, and interpersonal activities.

New technologies can enhance and enrich performances of understanding in many ways, including the following:

- Multimedia technologies allow learners to investigate new ideas and produce knowledge using a range of intelligences.

Sequence of Performances of Understanding

1. *Introductory performances* that connect to students' interests and initial levels of understanding

2. *Guided inquiry performances* that initially include more guidance from the teacher and gradually transfer more autonomy and responsibility to the learners—all the time requiring students to "go beyond the information given" and use their minds

3. *Culminating performances* that require learners to synthesize a range of understandings, perhaps encompassing different subject matters, developed over a considerable period of time but still focused on the target goals

---

- Many software programs incorporate features that assist learners with special needs due to poor motor coordination, diminished sight or hearing, or particular cognitive disabilities that interfere with learning such as short-attention span, memory problems, or difficulty analyzing visual or verbal material.

- Modeling software and simulations can make abstract concepts visible in ways that allow students to understand elusive ideas through actively experimenting, manipulating variables, and observing the dynamic interaction of elements in a system.

- Word processors, digital audio and video technologies, and tools for creating Web sites allow students to express their understanding in a rich variety of media. These technologies also capture student work in forms that can be easily revised, combined, and distributed.

Such tools support collaboration and peer learning in ways that are cumbersome or impossible with the traditional tools used in schools.

Just as new technologies can enhance performances of understanding, this element of the Teaching for Understanding framework can help teachers design ways for learners to gain the most educational benefit from their use of new technologies. Sometimes the allure of open-ended software, hypermedia, and networked technology is so enticing that both teachers and students become entranced with the possibilities for using these tools and lose sight of their own target goals. The criteria for performances of understanding remind teachers to guide learners' work with new technologies in ways that develop and demonstrate understanding of goals, require students to stretch their minds, and engage a rich range of intelligences.

Sometimes the allure of open-ended software, hypermedia, and networked technology is so enticing that both teachers and students become entranced with the possibilities for using these tools and lose sight of their target goals.

Case Study: A Sense of Caring through iEARN Global Art Projects

One of Kristi's enduring goals is to engage her students in collaborating with peers around the world to foster their understanding and appreciation of multicultural perspectives. She likes to promote her core academic goals, including literacy and social studies, through engaging her young students in authentic projects that involve them in taking care of their local and global community. Her interest in the iEARN Global Art project began during the first iEARN International Teachers Conference held in Patagonia, Argentina, in 1994. Several teachers at the conference were interested in the cross-cultural research of Emilia Ferreiro and Ana Teberosky, who looked at the literacy development of young children in Europe and Brazil.[3] Their collaborative research showed that, across cultures, children's drawings represented their first demonstrations of understanding story schema. Using this research as a launch pad, several iEARN elementary teachers decided to use children's original artwork as the basis for developing their writing and promoting global communication on a curricular topic of international importance. Eventually, they decided to focus the Global Art Project on the topic, "A Sense of Caring."

Understanding Goals

Kristi's understanding goals for this project encompassed two parallel strands:

1. Understanding a sense of caring through integrated social studies and service learning:
    A. locally in the classroom and community
    B. globally through collaborative communication with global peers
2. Understanding the purposes and processes of writing:
    A. expressing personal experience visually, verbally, and in writing
    B. exchanging ideas with local and global audiences through art and text products

Because Kristi addressed many of her key goals about both social studies and literacies in this project, she could afford to devote considerable time to a sequence of performances of understanding with her students. They included building up from introductory performances, to several guided inquiry performances, and then a culminating performance. Throughout this work, Kristi and the students used new technologies in a variety of ways to support students in developing and demonstrating their understanding.

Introductory Performances

To help students connect the topic, "A Sense of Caring," to their own experience, Kristi involved her students in conversations about caring with the whole class, in small groups, and in individual conversations with classroom peers and in conference with the teacher. In these talks, Kristi asked students to

- Describe examples from their own experiences about acts of caring by themselves and by others in their school, home, and community
- Give examples of receiving caring actions from others at school, home, and community
- Describe effects of giving and receiving caring
- Describe the characteristics and qualities of caring behaviors

Through these conversations, students developed their capacity to describe their personal experiences of caring and to listen to their peers and learn from them to construct a more complex understanding of varied perspectives on caring. These introductory activities allowed students to explore and articulate their own experience as a first step in becoming engaged with the curricular topic.

Guided Inquiry Performances

Building on this foundation of children's personal experience, Kristi orchestrated a series of performances that progressively deepened her students' understanding of a sense of caring and their capacity to express this understanding both in art and in writing.

Listening to, Reading, and Discussing Children's Literature About Caring

Using books that Kristi selected from the school and the classroom library, students listened to, read, and discussed stories with themes and examples of caring. These activities included whole-class and small-group lessons, with opportunities for individual reading followed by discussion with a peer, a small group, or the whole class. Through these discussions, the children made connections between themes and examples of caring in the books and their own first-hand experiences.

Drafting Individual Artwork

Students created an original pencil sketch to illustrate an experience depicting some characteristics and qualities of caring. They talked about their drawing with peers and in a teacher conference, including concepts about caring and ideas for revising the sketch to include more information, for example, details in the background, objects, and people to convey their thoughts more richly. Students revised their sketch to include more detail and then used the sketch as a model for creating artwork with crayon and watercolor. Students verbally presented the content and ideas of their artwork to classroom peers, answered questions from peers, and listened to their comments and suggestions for writing about the artwork.

Conferences with peers and the teacher helped students understand how they needed to embellish their verbal and visual depictions in order to represent and communicate their ideas effectively.

Through these experiences, students developed their capacity to represent their experience both verbally and through images. Conferences with peers and the teacher helped students understand how they needed to embellish their verbal and visual depictions in order to represent and communicate their ideas effectively. Exchanging their work and reflections with classmates also helped students appreciate how understanding develops through dialogue about diverse experiences and perspectives on a topic of shared interest.

Building on the research of Ferreiro and Teberosky, this art process produced a visual representation of ideas that Kristi portrayed to her students as parallel to the writing process (sketch = pre-write and draft; conference with teacher = planning for revising and editing; final artwork from sketch = revising, editing, and publishing). She described both art and writing as ways of putting students' ideas and experiences into a "document" to hold, share, revisit, and work with ideas.

Writing About Caring

To extend their understanding of caring, students created a written document based on their artwork. After trying this project once, Kristi realized that the process of developing students' written products could serve multiple purposes related to her literacy goals. Chapter Six includes a full description of Kristi's use of word processors through a structured progression of performances and assessments to help students elaborate and refine their written products.

Kristi realized that the process of developing students' written products could serve multiple purposes.

Students read their finished documents to the class and responded to questions from peers and the teacher. Students then shared both their artwork and writing with parents at conference time as part of their portfolio presentation. In each of these conversations—with peers, teacher, and parents—students developed and demonstrated both their understanding of a sense of caring and more metacognitive understanding of how they developed and extended their ideas through conversing, drafting artwork and written texts, reviewing and revising drafts, and sharing finished products in exchanges with a thoughtful audience.

Creating the Class Web Site

In recent years, the final performance of this phase included creating a class Web site where all the students' artwork was displayed, along with the accompanying text that students had written. For this Web site, Kristi took digital images of the students' artwork and loaded them onto the computer in preparation for students to work with the images on the Web site. Students used Graphic Converter™ and

Dreamweaver™ software to edit digital images and to create a class Web site that incorporated images of students and their artwork, written products, and tables. Kristi's goal in this phase was not for students to learn how to take digital images (as it was in the Quilt Math project described in Chapter Four) but to learn how to edit digital images for the purpose of creating a Web site to communicate about caring with global peers. As part of their lessons on Web site development, Kristi modeled the process of creating hyperlinks so students understood this concept in hypermedia and developed beginning understandings of Web addresses (URLs). Although Kristi's young students usually did not enter and work with the URL links, they understood the concept that URLs represent Web addresses, just as their street addresses represent the buildings where they live.

Culminating Performances

The culminating round of performances built on all the knowledge, skills, and values that students developed through the talking, drawing, writing, and sharing of their earlier work. In this phase, students extended their understanding of caring and their ways of acting on this understanding through exchanges with a global community of peers and by providing care in their local community.

Exchanging Artwork and Writing with Global Peers

Kristi collaborated with other teachers around the world on connecting their students through iEARN. When they first began working together in 1994, e-mail was solely text-based, so iEARN classes used e-mail to exchange written work but sent children's artwork through "snail mail." As the Web developed and attaching both text and image files to e-mail become commonplace, the iEARN classes increasingly used networked technologies to exchange all forms of student work and to promote global collaboration on this project. In recent years, iEARN classes involved in this project created their own class Web sites, where they posted digital images of students' artwork along with the associated texts that the students produced and video of their projects in the local community.

The culminating round of performances built on all the knowledge, skills, and values that students developed through the talking, drawing, writing, and sharing of their earlier work.

As a first step in exchanging students' artwork and texts with peers, students involved in the iEARN project explored their counterparts' class Web sites. The artwork was usually relatively easy to interpret, but because students wrote the texts in their own mother tongue and few of them could read other languages, the texts were often incomprehensible. Thus began students' appreciation for the diversity of languages in the world, the challenges of communicating without a shared understanding of words, and the ways visual arts can bring meaning to words of a language that is unfamiliar to students. Sometimes schools were able to provide a translation of their students' work into another language, most often English. Teachers also looked for local community people to come into school and translate and read some of the texts in foreign languages. Kristi's class used globes, atlases, and wall maps to locate and read names of cities, regions, countries, and continents of participating global peers.

Analyzing, Discussing, and Responding to a Global Art Show

Kristi's students viewed the Global Art Show and discussed the work of global peers in individual conversation with the teacher, with a peer partner, in small groups, and as a whole class. The students talked about ideas of caring displayed in the works, including commonalities and diversities among the artwork of global peers and their own class. Students responded to one another's comments with questions to clarify ideas and identify variations in the styles, materials, and medium of artistic work, as well as the content and forms of the writing (including a comparison of words, spellings, and style of script among the diversity of world languages represented in the show—for example, Russian, Spanish, Zuni, Chinese, and English).

Using the writing process described in Chapter Six, students developed e-mail messages individually or in small groups to respond to their global peers' artwork and writing about caring. Over several rounds of exchanges, students came to see commonalities in the themes in students' work from around the world as evidence affirming that other kids were like them. They also appreciated the diversities and contrasts as an opportunity to understand ideas about caring and about approaches to artwork that they hadn't previously imagined. Some of the best examples of children's e-mail writing can be found at http://www.psd267.wednet.edu/~kfranz/SchoolYear0102/socialstudies0102/GAPcaring02/GAP0102.html.

Over several rounds of exchanging emails, students came to see commonalities in the themes in students' work from around the world.

The immediacy of exchanging e-mail developed students' feeling of continuity and cohesiveness in sharing ideas globally in ways not possible without the Internet. The class Web sites, including a QuickTime® video example of a school in Argentina (http://www.smt.edu.ar/promociones/promocion2014/2002de2014/asenseofcaring/), also provided immediate access to global art and opportunities for students to share the project with families on computers at home or at the public library.

Sharing and Caring in the Community

As a final round of culminating performances of understanding, Kristi's students undertook a sequence of presentations and acts of caring. They began by sharing within their school and then extended their scope to the local community beyond the school. As a first step in this phase, Kristi's class invited other classes in the school, parents, the principal, school staff, and school district administrators to visit the Global Art Show. Students gave tours of the show to these visitors and shared their reflections about the caring ideas represented in the artwork and writing. Their presentations included reflections on (1) how they developed local-to-global understanding about caring, world languages, and world geography through collaboration with classmates and global peers and (2) how they developed their art and writing skills in the process of doing this project.

Based on all the ideas from global peers and the responses from their local community, the class decided how to act on their understandings of caring.

Next, the Global Art Show was displayed in the community public library. The class invited the community to view the show and write comments in a notebook provided at the show. Students took their families and friends to the show at the library. Back in their classroom, Kristi's students read and discussed the comments of visitors to the show in order to further extend their analysis of ideas about caring.

Based on all the ideas from global peers and the responses from their local community, the class decided how to act on their understandings of caring. They considered places in their local community and locations around the country and world where children's and adults' lives had been disrupted by natural disasters or periods of conflict. For example, global peers' ideas about caring for family elders inspired Kristi's class to conduct their Elders' project (http://www.psd267.wednet.edu/~kfranz/SocialStudies/socialstudies200001/iearnelders.html). Another year, Kristi's students learned about the impact of hurricanes on some of their peers in Puerto Rico and responded by creating Comfort Quilts (http://www.psd267.wednet.edu/~kfranz/SocialStudies/socialstudies200001/comfortquilts0001/comfortquilt0001.html and http://www.psd267.wednet.edu/~kfranz/SchoolYear0102/socialstudies0102/comfortqu.html).

Key Features of Performances of Understanding

All the essential features and characteristics of good performances of understanding can be seen in Kristi's Sense of Caring unit.

First, students' major learning activities *develop and demonstrate understanding of target goals.* One strand of Kristi's goals focused on various social studies topics, including understanding a sense of caring from personal, local, and global perspectives. A second strand focused on understanding how to develop and express ideas through dialogue with a range of audiences, engaging in structured collaboration with local and global peers, and developing and exchanging both artwork and writing with colleagues. All the learning performances directly and explicitly helped students develop and demonstrate their understanding of these goals.

Second, each performance required active learning and creative thinking to *stretch learners' minds.* The work was not simply a matter of repeating what students had been told. Instead, students generated their own ideas about concepts, actively critiqued what they read in relation to target concepts, created their own artwork and texts to express their ideas, and designed ways to demonstrate their understanding of the important ideas they had learned. These performances required students actively to compare ideas from multiple perspectives, to reflect on their own responses, and to make reasoned choices.

Third, students' performances built up understanding from *introductory performances,* through *guided inquiry,* to *culminating performances.* Kristi began by asking her students to talk about their own experiences with giving and receiving care. David Hawkins,[4] who believed strongly that effective learning had to begin with the learners' own interests and direct experience, advocated messing about with ideas as a good start. This kind of performance allows students a great deal of latitude in exploring new conceptual territory according to their own inclinations. As a next phase, Kristi structured a sequence of *guided inquiry performances* that allowed students to explore ideas of caring first through visual art—a more accessible form of thinking and expression—and then through writing.

The processes of creating both artwork and written work followed the same sequence of steps: brainstorming ideas, drafting or sketching, exchanging drafts, and conferencing around editing, revising, and creating and publishing a final product. During these middle-phase performances, the teacher provided structured guidance and support as necessary until students were able to perform well without such assistance. Finally, Kristi designed *culminating performances* that allowed her students to develop and demonstrate their understanding in ways that synthesized multiple strands of learning and required more independence from the students. In the first round, the culminating performance was presenting individual artwork and text. In the second round, the culminating performances included presenting the Global Art Show to multiple audiences and carrying out a community act of caring that demonstrated what students had learned throughout the Sense of Caring project. Such a series of performances allows a teacher to differentiate instruction to meet the needs of students with different interests, skills, and needs. These performances offered both support and opportunities for advanced work so that all students received the help they needed while working toward high standards of achievement.

Fourth, students' work in the Sense of Caring Unit included a *rich variety of entry points and multiple intelligences.* Howard Gardner[5] notes that humans approach learning and experience through a range of entry points, including narrative, dance, visual arts, music, and philosophy. In many respects, these entry points take advantage of different mental capacities to think and express ideas, which (as mentioned earlier) Gardner has dubbed multiple intelligences. Over several years of research on this topic, Gardner has identified seven distinct intelligences: mathematical-symbolic,

TIP

An effective set of performances of understanding engages a rich variety of entry points and multiple intelligences.

The Shakespeare Scene Exchange[1]

Contemporary students often find the unfamiliar language and contexts of Shakespeare challenging and have difficulty connecting with the characters and universal ideas in Shakespeare's plays. In the Shakespeare Scene Exchange project, seventh-grade students prepare and perform scenes from *A Midsummer Night's Dream* to learn about poetic and dramatic forms, character interpretation, and literary themes. Suburban middle school students and urban high school students work both in parallel and collaboratively, using e-mail and videos to compare interpretations of character and scene development.

To guide students toward an understanding of the artistry and meaning of the play, the teacher in each school engages students in a sequence of introductory activities, including lessons on vocabulary, poetic forms, character development, staging, and other dramatic and literary elements. Students then are organized into performance "troupes," who work through guided exercises to understand the language and characters, as well as to stage and rehearse scenes. Each class articulates assessment criteria for their performances, which they use to analyze scenes after each rehearsal. Rehearsals and critiques are videotaped to support discussions and to help students improve their individual performances. Video supports understanding by allowing students to view their performances from the "outside" and by providing specific evidence for learners' self-assessments and critiques, as well as the teacher's assessment of progress toward the unit's understanding goals. As students develop characters and scenes, the urban and suburban classes communicate via e-mail to compare similarities and differences of interpretation.

As a culminating performance, students present their scenes in full production before an audience of parents, teachers, and peers. Final performances are videotaped and exchanged with the partner school. Students write and share critical reviews and reflection papers comparing how the choices they and their partners made affected the meaning and presentation of the text in performance.

Through e-mail and video exchanges between schools, students observe how children of different ages and communities interpret and play the scenes of Shakespeare. By collaborating beyond the context of their own classroom, students gain a deeper understanding of universal relevance and varied interpretations of Shakespeare's work.

Resources

Folger Library Teaching Shakespeare
http://www.folger.edu/education/teaching.htm

Teachers First Guide to Shakespeare's Plays
http://www.teachersfirst.com/shakespr.shtml#midsummer

Surfing with the Bard
http://www.ulen.com/shakespeare/

Online Resources for Teaching Shakespeare
http://www.ericdigests.org/2003–3/online.htm

The Complete Works of William Shakespeare
http://the-tech.mit.edu/Shakespeare/works.html

Mr. William Shakespeare and the Internet
http://shakespeare.palomar.edu/

[1]This vignette is based on the work of Lisa McDonagh.

Multiple Intelligences

- Mathematical—symbolic
- Kinesthetic
- Verbal
- Visual

- Auditory
- Interpersonal
- Intrapersonal

---

kinesthetic, verbal, visual, auditory, interpersonal, and intrapersonal. Because students vary in their interests and their profile of intelligences, a rich range of performances allows more students to become engaged and learn effectively. Most school activities heavily favor verbal learning through talking and writing, along with mathematical-symbolic intelligence. Kristi's project offered additional entry points and engaged other intelligences—artistic (visual) as well as narrative—along with multiple opportunities for intrapersonal reflection and interpersonal exchange and collaboration.

Using New Technologies to Enhance Performances of Understanding

In order to take best advantage of the potential educational leverage that new technologies offer, it is wise to consider the features of effective performances of understanding and ask, "How might new technologies help incorporate these features into designs for learning?" Approaching the use of technology in this way reminds educators to focus on improving the learning processes and outcomes, not just adding some attractive bells and whistles to the educational experience.

Kristi used multiple technologies in the Sense of Caring project—all designed to help students develop and demonstrate their understanding of one or both strands of goals: (1) social studies and (2) expression through artwork and writing. Her students used word processors to develop, compose, and revise their reflections and the questions they sent to global peers about their artwork and writing. Word processors helped the students review and revise their writing, taking account of the comments from their teacher, classmates, and global peers through several rounds of editing. (See Chapter Six for a fuller description of this writing process.)

Teachers and students in this project made use of e-mail to promote speedy communication among students in a range of locations across the United States and around the world. The immediacy of exchange afforded by e-mail allowed students to feel deeply connected to distant peers and helped them make coherent sense of the various messages they sent and received.

Networked technologies also supported the exchange of visual and video materials as easily as text. Furthermore, class Web sites offered on-line spaces to construct and share the Global Art Show. By comparison with snail mail, the on-line art show was much faster, accessible from many places at once, revisable to allow students to design displays that best expressed their understanding, and "linkable" so that the various artworks and texts of many different students could be connected in ways that were meaningful to the learners. In all these ways, networked technologies supported a rich variety of performances, including interpersonal exchange and collaboration.

Kristi's class also used language translation software to translate English text to Chinese for sharing their artwork and writing with children in hospitals in China. This software not only enabled expressions of care but also developed Kristi's students' appreciation of the multilingual world as a medium of expression. This goal was part of Kristi's overall strand of goals associated with the development of students' appreciation of literacy as a means of formulating, representing, and communicating understanding.

TIP

For teachers who think this project sounds too elaborate to be relevant to their own situation, it is worth recalling that Kristi began with a much less complex project.

Planning and Teaching Performances of Understanding with New Technologies

Kristi's project, A Sense of Caring, is an ambitious, mature realization of many years of work on teaching for understanding with new technologies.

There are many possible entry points from which teachers might begin developing their own ways of using new technologies to enrich their learners' performances of understanding.

Getting Started Toward Coherent Sequences of Performances of Understanding

The description of Kristi's project outlines two cycles of performances of understanding: the first develops students' individual capacities to express their understanding of a sense of caring, and the second engages them in sharing and demonstrating this sense in their communities. Teachers could easily do just one of these rounds, and Kristi recommends starting with the community activities.

There are many possible entry points from which teachers might begin developing their own ways of using new technologies.

Why start with community work? Kristi believes that the benefits of penetrating the classroom walls and connecting students with the world outside are immensely valuable in motivating and educating learners. Kristi says,

TIP

This process of sharing work with peers and other audiences outside the classroom can start anywhere that feels interesting and manageable to the teacher.

> The power of the new technologies lies in their use for sharing and communicating so that students can see their work reflected back in another voice. As they share

Elevator Physics[1]

One purpose for studying physics is to recognize principles that underlie the mechanics of the physical world and to apply them to solving increasingly complex problems. Because learning physics involves understanding the dynamics of forces that cannot be directly observed, students often have difficulty grasping core concepts. Simulations can support students' understanding by enabling them to visualize and reason about abstract ideas applied to real-world problems. Particularly when students develop simulations themselves, they engage in performances of understanding that involve higher-order thinking, such as explaining, predicting, hypothesizing, exemplifying, demonstrating, and explaining. New technologies like Macromedia's Flash® animation and the Java programming language make the process of creating simulations accessible to students. In this unit, university freshmen studying introductory physics and computer science in two separate courses design and develop interactive simulations that model physics principles. They understand Newton's laws of motion through constructing a simulation of the ascent and descent of an elevator.

The physics instructor begins engaging students in *introductory performances* by posing an open-ended problem: How can we develop a simulation of the motion of an elevator to make the operation of Newton's laws visible? Working in teams, students begin to analyze the problem based on what they already know. Some students estimate or measure the height of the building and each floor, ride the elevator many times, and use chronometers, graphing calculators, and spreadsheets to predict how long the elevator will take to accelerate and decelerate at each floor. Other students research the capacities of the motor (power, energy consumption) or estimate the weight of the elevator counterweight and the cabin. Students bring their data to class, compare each team's approach, and discuss the theories that shaped their approach to data collection. The discussion enables the instructor to analyze and address each team's current level of understanding and misconceptions about Newton's laws of motion.

Next, students work through a series of *guided inquiry performances* to learn how to apply mathematical concepts and notation to the elevator problem and to discover how the interaction of forces and velocities must be represented in a simulation. To familiarize students with simulation programming, the physics instructor provides templates and a self-directed tutorial for developing Flash® animations. Students engage in calculating, estimating, and experimenting to animate the simulation at a realistic scale. Through these performances of understanding, students became intensely engaged in learning both physics concepts and the logic of computer programs, while analyzing an authentic, accessible example. Meanwhile, the computer science class learns to write Java applets—interactive computer programs that can be accessed on a Web page.

As a *culminating performance,* students in the two courses work together to create the elevator simulation programmed in Java. First, the computer science students teach the physics students enough about Java programming so the physics students can develop a realistic specification. Then physics students teach the computer science students about Newton's laws by writing a detailed specification for the simulation. Through engaging in metacognitive reflection in order to communicate abstract concepts clearly to one another, students in both the physics and the computer courses develop their understanding.

Resources

The Physics Classroom
http://www.physicsclassroom.com/Default2.html

National Institute for Science Education
http://www.wcer.wisc.edu/nise/cl1/

National Science Education Standards	High School Level Physics Tutorial
http://stills.nap.edu/html/nses/	http://www.glenbrook.k12.il.us/gbssci/phys/
Physical Science Resource Center	Class/BBoard.html
http://psrc-online.org/	National Science Foundation Digital Library
	http://www.nsdl.org/

[1]This project is based on the work of Cesar Nunes at the University of Sao Paolo in Brazil.

their work with different audiences, they begin to see why and how they need to clarify their thinking and revise the way they are expressing their ideas. When students see a clear purpose for working on their writing, that generates the energy for putting in the effort to create work they want to share.

Some teachers begin the Global Art Project by sharing students' artwork without much focus on writing. Others do not submit their own students' artwork but rather engage their class in sending e-mail about the art that other classes have shared. Although Kristi orchestrated a complex process of helping her students produce both pictures and texts before they participated on-line, at first she began with a simpler process. Initially, she had students post their first-draft artwork. She discovered how eager her students became to revise their art and their written work in response to comments from peers. This discovery led her, gradually over several years, to devise the parallel processes of developing artwork and written texts through rounds of critique within her class before posting student work on-line.

Kristi also gradually revised the steps in the culminating performance, which led her students to take action in their local and global communities to express their sense of sharing. It is important to realize that an elaborate project like the Sense of Caring unit evolves over many years, as teachers clarify their goals and refine their designs for learning activities.

Getting Started on Collaborative Performances of Understanding

For both teachers and students, collaborative work is risky and may be frightening. The culture of most schools values polished performances, that is, performances by teachers who already know what they're doing and students who can produce the right answer fast. With these kinds of values, sharing early draft work and collaborating with others on innovative pilot projects may feel potentially embarrassing. What if I don't know what I'm doing? What if our work is not very good at first? What if I make mistakes? Both students and teachers may worry about these issues.

What if I don't know what I'm doing? What if our work is not very good at first? What if I make mistakes? Both students and teachers may worry about these issues.

Using new technologies and trying to teach for understanding are both endeavors that require learning on the job. It is not adaptive to try to figure out everything in advance before attempting either of these innovations, let alone doing both at once.

To do this work, one is better off embracing a tolerance for uncertainty that is not common in most schools. Consider that taking action despite not knowing exactly what to do is a sign of courage, not ignorance. Putting out work that is not yet polished is a way of learning with others through collaborative co-construction of knowledge. It often generates better work ultimately than any one person can produce alone. Making mistakes is a sign of willingness to risk learning and growing, not evidence of failure. Cultivating these kinds of beliefs in one's students is essential if they are to develop the courage to share early drafts in hopes of gathering suggestions that will help them deepen their understanding and improve their work.

Kristi believes that teachers are more likely to help students embrace these beliefs if the teachers themselves enact them. Such teachers are willing to ask for help when they don't know how to do something, game to "take a crack" or "make a stab" as a way to start when they don't know exactly what to do, and they don't mind acknowledging publicly that they are still learners. Teachers who are courageous learners themselves are more likely to foster their students' courage to learn. Furthermore, recent firsthand experience with the anxious feel-

---

Building a Culture that Supports Teaching for Understanding with New Technologies

Using new technologies and trying to teach for understanding are both endeavors that require learning on the job. It is not adaptive to try to figure out everything in advance.

- Taking action despite not knowing exactly what to do is a sign of courage, not ignorance.

- Putting out work that is not yet polished is a way of learning with others through collaborative co-construction of knowledge.

- Making mistakes is a sign of willingness to risk learning and growing, not evidence of failure.

- Ask for help when you don't know how to do something.

- Be a learner to model learning with your students.

- Firsthand experience with the anxious feelings of learning often makes teachers more sensitive and effective in helping their students.

ings of learning often makes teachers more sensitive and effective in helping their students.

Getting Started with On-Line Collaboration

Teachers who are not accustomed to collaborating within their classrooms may need to build up gradually to on-line collaboration. Even though Kristi became very familiar with on-line collaborative projects, the new students who entered her classroom every year usually were not. So she gradually helped them develop the skills and understandings they needed to participate effectively in the Global Art project.

As a first step, she conducted a collaborative exchange with another teacher in her own school. One of the kindergarten teachers became interested in a project focused on a sense of caring, so Kristi worked with her to foster exchanges of art-work among the students in their classes. This project allowed students to share their work with peers whom they could readily meet face-to-face. When meeting students from the other class on the playground or in other settings, they could talk about their shared project and begin to gain a sense of how peer exchanges contribute to understanding.

Next, Kristi arranged an on-line collaboration with a close colleague who taught in Seattle—across the state from Kristi's town. Her students had heard of Seattle, and many had been to visit there. So these students in Seattle seemed distant, but not too foreign. During this exchange, Kristi's students learned how to communicate on-line and how to express themselves clearly to peers who were from a very similar culture. These students were not likely to meet face-to-face, however, so they learned how to invest in working closely with students they would not see.

> **When she first worked with this trusted colleague, Kristi did not know how the lessons would play out.**

These initial pilot efforts also allowed Kristi to practice her skills in on-line curricular collaboration outside her own school building and district. When she first worked with this trusted colleague, Kristi did not know how the lessons would play out. Because she felt safe with her fellow teacher, she found it easy to talk with her, either in meetings or on the telephone, to think about how to revise their approach. Working with a close colleague gave Kristi an opportunity to build her confidence in collaborating on-line with colleagues in distant locations around the world. It also helped her develop skills with the technology and with the process of organizing her students' work on-line. Through these early pilot projects, she discovered how to create a culture with her students that valued innovating, taking risks, and working collaboratively. Devising pilot projects that are not such a stretch creates a form of safety net that makes subsequent, more ambitious on-line collaboration feel less risky.

**Build Teacher and Student On-Line Collaborations
for Teaching and Learning Gradually**

- In your own building
- In your geographic region or state
- In your country and across continents

Getting Started Using New Technologies

Learning to use new technologies and finding the resources and help to do so are also gradual processes. As Kristi began to see ways that computers could support her preferred ways of working with students, she sought out people who could help her learn. These included friends, parents of students, and some of the technology specialists in her school. Working with a knowledgeable adult turned out to be more effective for her than taking formal classes.

Soon, Kristi was the most technologically advanced teacher in her school, and she was asked to become the technology specialist for her building. She still did not feel that she was much of an expert but realized that she understood more than most of her colleagues, so she happily took on this role. In this position, Kristi was given more equipment and assistance than most teachers might receive so that she could learn how to use the technology in preparation for mentoring her colleagues. While accepting these additional resources, she continually looked for ways to share them and her own expertise with other teachers. This willingness to assist others was essential, both in sustaining a generous relationship with the other technology specialists and in preventing the kind of collegial resentment that sometimes grows around an innovative teacher.

Technology support people often devise formal rules for gaining access to their assistance so they will not be overwhelmed, but these procedures can seem impenetrable. Everyone, from the teachers just starting to use technology to the technical experts, has to work on being patient with one another.

Kristi also worked on maintaining a positive relationship with her principal and superintendent. As her students carried out innovative projects with new technologies, Kristi kept her administrators informed by showing them examples of student work. Their familiarity with the benefits of these projects meant that the administrators were well informed on how new technologies were effectively supporting teaching and learning. Therefore, they were prepared when they received calls about Kristi's work (for example, requests to film Kristi's classroom for a PBS documentary) or special requests for support that Kristi might make.

TIP

Using technology in a school is often more of a human relations challenge than a technical one. It requires building and maintaining good connections with peers, technology specialists, and administrators who allocate valuable resources like time, budget, and professional opportunities.

Extending Innovation with Colleagues

In many schools, teachers feel so stretched by meeting all the demands of their daily responsibilities that the prospect of adding extra innovative projects is daunting. They may resent pioneers like Kristi who are continually experimenting and exploring beyond the normal boundaries. "You have to be proactive in building a collaborative community that invites and honors each person's participation and contributions so that a culture of resentment doesn't develop," says Kristi. She gave a presentation to her fellow faculty about the Global Art Show project and invited all of them to participate. The school's art teacher had already offered to help students create their art projects during art class so that other teachers could use students' art projects to prompt writing assignments in their own classes. Kristi encouraged teachers to consider participating in the Global Art project in whatever way they wished. For example, they could just share their students' art, or use the art posted by other students and focus only on the writing process, or have young students dictate their stories to older students who would type them. "Whatever piece you want to do is OK. Having students write even a few sentences is fine—it doesn't have to be a whole essay. You can start with an appetizer or dessert before biting into a huge main course," she told her colleagues. Kristi also offered to help teachers with any concerns about using technology, engaging students in a process of writing, and meeting challenges.

Teachers feel so stretched by meeting all the demands of their daily responsibilities that the prospect of adding extra innovative projects is daunting.

By working hard to share her ideas, knowledge, and relationships, Kristi managed to involve many other students and teachers in the Global Art Show project. During several years, her entire school participated in the project, using varied approaches. Other years, only four or five other teachers opted to join. The important goal for Kristi was to help other teachers and students, as well as her own students, take advantage of new technologies to improve learning through collaboration and exchange.

Questions for Reflection

1. How might you design a sequence of performances that would develop and demonstrate your students' understanding of your target understanding goals? In order to build from where your students begin, consider planning an introductory or "messing about" performance, a series of "guided inquiry" activities that build students' capacities around target knowledge and skills, and a culminating performance in which students work more autonomously

to synthesize and apply what they have learned. Be sure that these performances require students to stretch their minds, not just rehearse routine knowledge and skills.

2. How might a new technology that is accessible and feasible to use in your setting help students develop and demonstrate understanding significantly better than they could do with traditional tools?

NOTES

1. David Perkins discusses performances of understanding at length in *Smart Schools: From Training Memories to Educating Minds* (New York: Free Press, 1992, pp. 75–79). Here he also cites Jerome S. Bruner's elegant phrase, "Going beyond the information given" from *Beyond the Information Given* (J. Anglin [ed.]. New York: Norton, 1973, pp. 468–479).

2. Gardner, H. *Frames of Mind: The Theory of Multiple Intelligences.* New York: Basic Books, 1983; *Intelligence Reframed: Multiple Intelligences for the 21st Century.* New York: Basic Books, 1999.

3. Ferreiro, E., and Teberosky, A. *Literacy Before Schooling* (trans. K. G. Castro, preface by Y. Goodman). Exeter, N.H.: Heinemann Educational Books, 1982.

4. Hawkins, D. *The Informed Vision: Essays on Learning and Human Nature.* New York: Agathon Press, 1974.

5. Gardner, H. *The Unschooled Mind: How Children Think and How Schools Should Teach.* New York: Basic Books, 1991.

Ongoing Assessment and New Technologies

More than fifty years after completing her formal schooling, my grandmother had recurring nightmares about confronting the final examination in a course that she never attended because she forgot she had enrolled in it. The fact that this classic anxiety dream afflicts many people attests to the role of assessment in most schools. High-stakes tests are used at the end of a course to rate and rank students' achievement. Often the assessment is based on tasks that have little to do with the actual goals the teacher hopes students will reach. Or the assessment may be based on a final product that the student labored mightily to produce but will never be seen by anyone except the teacher. Such assessments do little to advance students' understanding. Indeed, they often undermine students' confidence and diminish the motivation and determination that contribute to effective learning.

Assessments can significantly promote learning, however, if they are designed and conducted with this purpose in mind. Research has shown[1] that students who regard learning as an incremental process of developing understanding tend to learn better than those who see learning as "getting it" versus "not getting it." Students with an "incremental" view of learning tend to risk sharing draft work and to persist when they cannot immediately figure something out. In contrast, "entity" learners are more likely to give up in defeat if they don't immediately know "the answer" or see a way to solve a problem. Helping learners embrace a more incremental perspective enables them to view learning as a process and to regard themselves as capable of advancing their own learning through effort, practice, and revision.

> ## How Ongoing Assessment Supports Understanding
> - Students understand what quality work entails.
> - Peer collaboration helps students analyze and improve their own work.
> - Frequent checking along with feedback from multiple sources helps students see and accept multiple ways of improving their work.
> - Teachers gain a more complete picture of their students' understanding.

TIP

The most effective teachers conduct frequent assessments throughout the learning process, not just at the end.

The Teaching for Understanding project found that the most effective teachers conduct frequent assessments throughout the learning process, not just at the end. They develop clear assessment criteria, that is, qualities that define high-quality work, that focus directly on priority understanding goals. Teachers publicize these criteria openly with students so that students themselves can monitor their own work and participate in peer assessments. As students produce initial drafts, frequent assessments help students recognize the strengths of their work so far and the ways it could be better. Assessment includes not just rating the work but also providing suggestions for improvement.

Ongoing assessments that are conducted in relation to public criteria and that generate constructive recommendations directly support the development of understanding in several ways. First, students understand what quality work entails when they play an active part in conducting assessments, for instance by joining in the process of defining assessment criteria and by participating in self-assessments and peer reviews of draft products. When students themselves understand what constitutes high-quality work, they are much more likely to be able to produce such work. Second, students benefit from seeing how their fellow learners are approaching a lesson or project. Peer collaboration on assessment often helps them see how to analyze and improve their own work. Third, frequent assessments provided by a variety of people—fellow students, outside advisers, and the teacher, along with self-assessment—may help students see and accept multiple ways of improving their work. Finally, teachers who conduct varied ongoing assessments of student work gather a more complete picture of their students' understanding in time to design suitable interventions. When students receive constructive suggestions on early drafts and then have multiple opportunities to revise their work, the final product is likely to be much stronger than if the only assessment takes place after students have completed their work.

New technologies provide several advantages in conducting these kinds of ongoing assessments. Digital technologies, including audio and video recorders and computers, can capture student work in forms that are easy to review. Interactive workspaces and software with multiple windows can help to keep assessment guide-

---

Key Features of Ongoing Assessment

- Assessments are conducted throughout the learning process, not just at the end.
- Clear assessment criteria are developed that focus directly on priority understanding goals.
- Criteria are publicized so that students can monitor their own work and participate in peer assessments.
- Assessment includes not just rating the work but providing suggestions for improvement.
- Assessments are provided by a variety of people—fellow students, outside advisers, and the teacher, along with self-assessment—to help students see and accept multiple ways of improving their work.

---

lines in view and may even offer prompts and reminders as students work. In this way, technologies support the processes of analysis and reflection that are central to constructive assessment. When students' work is captured with digital technologies, revision is less burdensome because learners can change only the parts that need improvement instead of having to redo the entire product. Using networked technologies, students may post their work on-line where it can be readily reviewed and annotated by multiple advisers, including distant teachers and peers who cannot meet face-to-face. Technologies also provide easy means of preserving digital archives of student work. These may allow teachers and students to create individual portfolios to demonstrate and evaluate a student's progress over time.

Ongoing Assessment Supported with New Technologies

- Digital technologies, including audio and video recorders, as well as computers, can capture student work in forms that are easy to review.
- Interactive workspaces and software with multiple windows can help to keep assessment guidelines in view and may even offer prompts and reminders as students work.
- With products created using digital technologies, revision is less burdensome because learners can change only the parts that need improvement instead of having to redo the entire product.
- Using networked technologies, students may post their work on-line where it can be readily reviewed and annotated by multiple advisers, including distant teachers and peers who cannot meet face-to-face.

Such archives allow students to learn from assessing model examples created by former students, and they enable longitudinal research, such as the student research on a local water habitat described in Chapter Three.

Case Study: Writing for Understanding

One of Kristi's overarching goals is to develop her students' appreciation of learning through writing as a process of drafting to capture initial ideas, thoughtfully reviewing and discussing draft works to clarify and extend understanding, and revising to express more accurate, elaborate, and complete understanding. Her curriculum was saturated with opportunities for students to develop and communicate their ideas through the process of writing. She used ongoing assessment with a variety of technologies to support this process.

The Bird Print Writing Process

In the Sense of Caring project described in Chapter Five, Kristi's students used both artwork and written texts to develop, express, and share their understanding of care. Before posting their texts on-line to share with global peers, students worked through a systematic sequence of pre-write, draft, edit, revise, and publish, guided by ongoing assessment at each step.

Over several years of refining what became known as the Bird Print Writing process, Kristi searched for memorable titles to help students recall and appreciate key tasks and attitudes at each stage. The bird connection arose from Kristi's passion for watching birds and collecting prints of bird paintings, which fueled a study of birds she undertook with her students. Gradually, she identified a bird that her class was studying whose name and habits correlated in evocative ways with each stage of writing. The following summary is taken from a much fuller description of this work published on-line at http://www.psd267.wednet.edu/~kfranz/Literacy/birdprintwriting.htm.

Bird Print Writing Process

- *Puffin Pre-write.* List main ideas.
- *Swan Draft.* Write a sentence about each main idea.
- *Egret Edit.* Mark possible spelling and punctuation errors.
- *Owl Edit.* Correct spelling and punctuation.
- *Wren Revise.* Express ideas more fully.
- *Peregrine Publish.* Make final changes, format correctly, and publish.

Puffin Pre-write

In the Puffin Pre-write stage, students generated a list of words or short phrases to capture main ideas for a piece of writing. Kristi noted that this is somewhat like what puffins do as they flap their short wings, hovering around ocean cliffs looking for a place to land or dive for fish. Puffins live in colonies, rather than alone, and Kristi highlighted this habit to emphasize the community aspect of pre-writing. Students talked about the writing topic as a class or in small groups to help them begin to put their own ideas into words.

Then each student opened a new document with the word processor, typed the title of the project (for example, "Sense of Caring Puffin Pre-write") and their name and the date, and then typed a list of words and short phrases to indicate ideas they might want to include in their writing. They used the book spellings they knew and otherwise used "sound" spellings to record their ideas. Students put each word or phrase on a new line to facilitate adding more ideas later. When their list seemed complete, they used their computer's cut-and-paste feature to collect words about the same idea together but still in list form. This facilitated clustering sentences about the same idea together and forming them into paragraphs later in the process. The student saved this document on a disk or folder on the hard drive and printed out a hard copy to use in the assessment process. Finally, the student saved the file with a new name (for example, "Sense of Caring Swan Draft") to prepare for the next step.

Each student met with Kristi or a parent volunteer to discuss and assess his or her Puffin Pre-write document. Kristi used both a qualitative and a quantitative rubric to guide this conference. The qualitative assessment items included how well students verbally articulated sentence ideas for the words on their list, what the students said about the challenges and progress of their writing, and what strategies the teacher used to foster the students' understanding. On the quantitative assessment, Kristi and the student noted, for instance, the total number of words, number of sound spellings, number of book spellings, and number of pre-write words or phrases the student listed with no assistance from others (for example, peer, volunteer, teacher), with some assistance, and with modeling from the teacher. This assessment form was filed with the Puffin Pre-write document and became part of the student's writing portfolio to track progress within and across writing projects.

> *Each student met with Kristi or a parent volunteer to discuss and assess his or her Puffin Pre-write document. Kristi used both a qualitative and a quantitative rubric to guide this conference.*

Swan Draft

The swan gracefully floating across the water suggests the goal of writing a draft by letting sentences for each idea flow onto the page. The main goal of this stage of writing was to generate flowing sentences about each idea in the Puffin

Pre-write list, recognizing that correcting errors and polishing the writing would take place later.

In the Swan Draft document, the student typed one or more sentences under each word or phrase in the Puffin Pre-write list. The list of words helped students work with the ideas they generated and discussed during the pre-write process. By drafting sentences on the computer rather than on paper, students could type as much as they wished; they were not limited to the space around the word on the printed page. Students who lacked fluent keyboarding skills interspersed keyboarding practice with other computer-based steps of the writing process. When they finished their Swan Draft, students saved the file with double spaces between lines and printed a paper copy to use in their conference. They also saved this document with a new name—"Egret Edit"—to set the stage for the next phase of work.

Each student met with Kristi or a classroom volunteer to discuss and assess the Swan Draft. The quantitative assessment for this stage dealt with content, sentences, and conventions, indicating descriptive words to communicate ideas around the pre-write list, the number of complete and incomplete sentences, and appropriate use of capital letters and punctuation. The qualitative assessment included noting the student's metacognition or reflection about his or her own writing challenges and progress.

Revising was a more challenging task, as students often didn't recognize any need for changes. So Kristi supported editing in two phases.

Egret Edit and Owl Edit

Kristi found that primary students understood the process of reviewing their own writing more readily if they began by editing for spelling and punctuation. Revising was a more challenging task, as students often didn't recognize any need for changes. So Kristi supported editing in two phases. The first phase is named for the egret, who has keen eyes to look closely at its habitat to find food and who strikes quickly with its beak to catch prey. During the Egret Edit stage, students looked over their own Swan Draft (on the computer, if possible) and underlined words they thought might be sound spellings that needed to be corrected to book spellings. They also looked for places where capitalization and punctuation might be needed. They saved this document as the Egret Edit text and then saved it again as their Owl Edit document.

During the Egret Edit conference with Kristi, the assessment sheet reminded students to tally the number of sound spellings they had underlined and the number of additional sound spellings the teacher underlined. They also tallied the number of accurate book spellings in the document, along with the number of capital letters and punctuation marks the student underlined. Kristi praised progress on correctly identifying sound spellings, as well as success on producing book spellings.

Many of Kristi's students became able to complete the Puffin Pre-write and Egret Edit effectively and independently.

Kristi or a parent volunteer worked intensively with students who needed more help. Kristi devoted years to cultivating the involvement of various adults (parent volunteers, student teachers, classroom aides, and Title I teachers), as well as the students themselves, in the assessment process. More information about this process and the way computers supported the work of small groups may be found in subsequent sections of this chapter and in Chapter Seven.

Kristi chose the owl to represent the second phase of editing because of the owl's keen eyesight for looking carefully in many directions and because this bird is a symbol for wisdom and knowledge. During the Owl Edit phase, students looked up the book spellings for words, wrote the correct spellings above the underlined words on the hard copy of their Egret Edit document, and then used the paper copy as a guide for correcting their file on the computer. Students might work together at this stage so that those who were more proficient with the computer and dictionary skills helped the others find and correct both spelling and punctuation errors. Kristi's conference with students at the end of the Owl Edit phase was guided by an assessment form indicating the number of book spellings the student located and corrected independently, with help from a peer, or with assistance from the teacher.

Wren Revise

The wren hops from branch to branch, reminding students of the way they can "cut and paste" to revise their work. It also has a beautiful lyrical song that reminds students to make their words and sentences "flow together like music." During the Wren Revise phase, students brought the paper copy from their Owl Edit work to a conference with the teacher. The student read the written document aloud, listening for missing words or incomplete sentences. Kristi also read the piece to the child to help the student recognize places where changes might be needed to make the piece communicate more clearly. Students wrote revisions on the paper copy. During this conference, Kristi also guided the students toward creating paragraphs. This process was aided by the earlier work of clustering words about the same idea during Puffin Pre-write.

Sometimes, Wren Revise conferences involved small groups of children so that they could learn from listening to and commenting on one another's revising efforts. Students often collaborated with a peer on Wren Revise work while they waited to meet with the teacher. After completing the teacher conference, the student entered final corrections and revisions into the word processing file, saved the file, and then re-saved a copy as "Peregrine Publish" and printed a paper copy.

TIP

The assessment conferences and forms provided clear evidence about each child's progress on reading and writing vocabulary, spelling, and punctuation, and they provided a basis for setting individual goals and work plans to develop each student's literacy.

Kristi also read the piece to the child to help the student recognize places where changes might be needed to make the piece communicate more clearly.

Peregrine Publish

The peregrine falcon is a fast-flying bird, which is why Kristi chose it to represent the final stage of the writing process when students posted their text to e-mail or a Web site so that it was speedily communicated to their audience. During the Peregrine Publish conference, the student read the document aloud, which Kristi used as an opportunity to assess reading fluency and expression, as well as the content of the document. The student also reviewed the document to identify any final corrections and to organize formatting for publication. Once the student made these final changes, the document was ready to be published and shared via e-mail or the class Web site.

Supporting Peer Assessments

The texts that Kristi's students developed through the Bird Print Writing process often became part of published products. As described in Chapter Five, the students' writing about a sense of caring became part of a Global Art Show, developed through exchanging work with other classes around the world. Both artwork and texts that were captured in digital formats became part of the classroom Web site. The original artwork created by Kristi's students, along with some digital reproductions of work from their global peers and their written texts, became part of an art show posted in the school. Publishing their pictures and texts for an authentic audience inspired Kristi's students to develop and polish their work. The exchange of ideas with peers around the world also led to further rounds of assessment, feedback, and revision. Once students had published their artwork and texts to the Global Art Show on-line and as an art show in their school building, they participated in a structured process of sending and receiving comments about their work. Networked technologies, such as the Web and e-mail, facilitated this kind of peer assessment and exchange. But Kristi learned that both students and teachers involved in global exchanges needed specific guidance and practice in order to take full advantage of the opportunities that communication technologies offer. As illustrated by the Bird Print Writing process, Kristi explicitly taught her students how to be constructive critics of their own work.

As Kristi studied her students' reaction to these activities, she came to understand that part of the challenge in editing and revising arose because students identified their work as part of themselves. Editing their work required students to separate themselves from their products, to gain more objective perspective on the work, to recognize that what they thought and intended did not always come through in the written product, and to acknowledge aspects of their written text that needed improvement. Kristi described this issue as a matter of being able to shift from being the *author* to the *audience* of a text. One reason she structured a two-step editing process *before* the revision process was to give students practice in the process of making this shift from author to audience. In the first round of editing—the Egret Edit phase—students looked at spellings and punctuation conventions that are more easily identified by young writers. After this editing experience, students were more able to tackle the second round—Owl Edit—when they engaged in the more complex challenge of recognizing syntactical shortcomings in their text, devising ways to improve the text, and incorporating revisions into the document.

Through these mini-lessons and the modeling of constructive feedback that students saw during the multiple conferences with adults in the Bird Print Writing process, Kristi's students developed skills for assessing their own and their classmates' work.

Kristi built on these skills in teaching her students how to participate effectively in ongoing assessment with global peers during the Global Art Show project online. In preparing her students to respond to the work of their global peers, Kristi engaged the whole class in a conversation about what their e-mail messages should include, and together they devised an "e-mail content rubric." They decided their e-mail to global peers should include

TIP

To develop students' capacity for being effective assessors, Kristi broke the process into subcomponents and devised mini-lessons on key assessment tasks.

- Acknowledgment and appreciation for receipt of the art and the written work

- Comments and compliments about the artistic and the writing effort evidenced in the work, with specific examples

- Comments on the ideas about caring communicated in the artwork and writing

- Comments about connections between the ideas about caring that were communicated in the artwork and writing of global peers, as well as their own experiences

- Thoughtful questions to extend communication on caring

By engaging her students in developing this rubric, Kristi helped them clarify and understand the purpose of their e-mail exchanges with global peers. The rubric then served as a guide for students to use as they planned, drafted, reviewed, and revised the messages they developed to post on e-mail. And the rubric reminded them how to be constructive providers of feedback to their peers.

Key Features of Ongoing Assessment

Kristi's work with her learners in the Bird Print Writing process illustrates important aspects of ongoing assessment and ways of using new technologies to support them. Ongoing assessment, as described in the Teaching for Understanding framework, includes some key features that differ from traditional assessment processes in schools in several significant ways.

Understanding Triangle Properties[1]

It is common for students to reproduce geometric proofs through memorization without confidently understanding the underlying properties and relationships of objects in a plane. This high school geometry unit involves students in continuously assessing their own understanding as they work cooperatively with their teacher and one another through cycles of conjecture, experimentation, discussion, and application to real-world problems to deepen their understanding of the properties of triangles.

Each class for this unit begins with a warm-up problem from a real-world situation to gauge students' practical sense of triangles, prior knowledge, and reasoning processes. In cooperative groups, students then work through a series of guided lessons prepared by their teacher that address types of triangles, triangle sums, triangle inequality, side-angle inequality, and similar triangles. For each lesson, students construct multiple triangles, gather data, and reason together about one another's constructions as they formulate, test, and compare their ideas, and pose arguments to the class explaining their reasoning.

As students conduct triangle experiments, they use Geometer's Sketchpad—a software tool to enable the dynamic exploration of mathematical ideas. Sketchpad supports learners' inquiry and self-assessment by allowing them to visualize and analyze a problem in order to frame and test conjectures. Students can construct an object, then explore its properties by measuring, dragging, reshaping, and comparing it to other objects. The software preserves the mathematical relationships of objects, allowing examination of a set of similar cases in order to observe patterns and make generalizations. It allows learners to record and review their work, identify their own targets of difficulty, and discover alternative approaches to solving problems. The software's accuracy of measurement and instant results allow students to test their conjectures quickly, freeing them from the tedium of construction in order to focus on reasoning and understanding.

By providing detailed, manipulable examples of students' constructions, Sketchpad supports work within cooperative groups, as well as class discussion about the patterns and structures that lead to an understanding of geometric proof. Because Sketchpad records each iteration of students' work, it also supports the teacher's assessment by providing a clear representation of students' misconceptions and evolving understanding.

Resources

The Math Forum Classroom Resources for Geometer's Sketchpad
http://mathforum.org/dynamic/sketchpad.links.html

Key Curriculum Press (publisher of Geometer's Sketchpad)
http://www.keypress.com/sketchpad/

Geometry and Proof
http://www.terc.edu/investigations/relevant/html/geometry.html

Selected Resources for Teaching Geometry
http://www.questia.com/popularSearches/teaching_geometry.jsp

National Council of Teachers of Mathematics (NCTM) Geometry Standards
http://standards.nctm.org/document/chapter3/geom.htm

---

[1]This vignette is based on the work of Mary Teixiera at TechBoston Academy in Boston.

Key Features of Ongoing Assessment

- Criteria for assessment are related to understanding goals.
- Basis for assessments is made explicit and public.
- Assessments are conducted frequently and generate suggestions for improving work.
- Assessment comes from multiple sources.

First, the criteria for assessment are *related to understanding goals*. Many assessments in schools consist of multiple-choice or short-answer tests that do not fully measure the goals that teachers have for their students. Because understanding goals are often ambitious and complex, teachers sometimes need to be creative in devising assessments that truly focus on these priorities. In the Bird Print Writing project, one key goal was for students to understand how to express themselves in writing clearly, fully, and appropriately for particular audiences. A second goal was to understand writing as a process of thinking and communicating. In contrast with these goals, many students believe that writing is simply a means of capturing their thoughts after they have completed their thinking. From this perspective, the first draft is the final product, and revising one's work feels more like fixing a mistake than improving on partially completed work. A third goal of the Bird Print Writing process was to understand how to guide and monitor oneself through the writing process. The rubrics that Kristi devised for guiding assessment at each stage of the process addressed these ambitious goals, as well as the more basic skills of using correct spelling, punctuation, and grammar.

A second feature of effective ongoing assessments is that the *basis for assessments is made explicit and public,* often in collaboration with students themselves. When students understand the assessment requirements, they are in a better position to work toward meeting the criteria. Kristi's students worked with her to develop the e-mail rubric for their messages to Global Art Show project peers. Then the rubric served as a guide for their work, as well as a basis for assessing their final messages.

Third, assessments are *conducted frequently* as students develop their work and the assessments *generate suggestions for improving* the work. This process, which looks more like coaching than evaluating, blends monitoring progress with supporting students' understanding. Assessment may be done informally and be embedded within learning activities, as when Kristi observed her students reading their draft texts to determine whether they recognized where their text was incomplete. It may also be more formal, as when Kristi completed a formal rubric with her students during their conferences at each stage of the Bird Print Writing process. And assessments may be conducted with groups as well as individual students. In all cases, an abiding purpose is to promote progress, not simply to monitor or evaluate achievement.

Fourth, assessment comes from *multiple sources,* including fellow students, self-assessment, and outside audiences, as well as the teacher. Students may need special help in order to become effective providers of constructive assessments for themselves and fellow learners. Kristi devised mini-lessons to develop assessment skills in her students. She believed that applying metacognitive habits of mind (for example, monitoring one's own process and progress as a learner) and understanding how to analyze the qualities of a piece of work) are important understanding goals. Participating in the peer assessment process also helps students learn from studying their classmates' work. Other contributors to assessment who understand the assessment criteria, such as the parent volunteers and the collaborating teachers who worked with Kristi's students, also provided valuable insights and assistance to learners.

How New Technologies Enhance Ongoing Assessment

One of the most valuable contributions of new technologies to ongoing assessment is that they bring (virtually) into the classroom authentic audiences who inspire learners to polish and improve their work. When students submit their work only to their teacher and expect to learn little more than the grade or score they earned, they are less likely to be incremental learners who invest in revising their products. Kristi's students, who knew they would be using e-mail and the Web to share their work with peers and other authentic audiences beyond their classroom, were motivated to assess their own work with a critical eye and to make their products as accurate and compelling as they could.

Kristi's students captured each stage of their work during the Bird Print Writing process and stored it in a digital portfolio that documented their progress. They also printed a hard copy of each phase of work in a format that facilitated assessment and review. Capturing work in this way helps both the student and the teacher recognize progress, as well as areas of weakness or confusion, while there is still time to make improvements. Kristi found that students who were able to inspect their work from the perspective of an "audience" rather than an "author" were able to notice and willing to correct problems that they might otherwise have overlooked. Documenting their work in concrete forms (for example, printing out drafts or seeing their artwork and texts prepared as a product to share with an external audience) helped students regard their output from the perspective of an audience. This was an important component in motivating students to critique and improve their work.

Digitized products can then be easily revised, eliminating some of the drudgery of revision. By creating their writing with a word processor rather than by

TIP

New technologies support the process of capturing student work in forms that can easily be reviewed and revised with assessment criteria in mind.

handwriting, Kristi's students had legible documents to review throughout the process from pre-writing to publishing. As students worked through the Bird Print Writing process, they identified the places where their text needed improvement, decided on the revisions they wished to make, and then corrected only the portions that need to be changed. By comparison with the old technologies of smudging out a mistake with a pencil erasure or fiddling with white cover-up liquid on a typewritten page, word processors offer obvious advantages. In addition, they permit a revised work to look just as good as the original—something that may encourage learners who would otherwise feel embarrassed by "mistakes."

Kristi found that students who were able to inspect their work from the perspective of an "audience" rather than an "author" were able to notice and willing to correct problems that they might otherwise have overlooked.

Networked technologies help to engage a range of participants in the assessment process. Learners often benefit from exchanging their work and their assessment feedback with fellow learners who share some of their concerns and their own level of beginning understandings. Kristi's students used both e-mail and the Web regularly to exchange work with global peers and to foster peer feedback. Kristi also used networked technologies, including Internet-based videoconferences, to permit her students to present their work to authentic audiences, such as the local city council, youth forums with other schools, and international conferences on education. Such presentations may constitute the most authentic form of assessment when the goal of learning is to develop learners' capacity to exercise a positive influence in the world.

Planning and Teaching Ongoing Assessment with New Technologies

Ongoing assessment challenges many of the usual values and patterns in schools. It focuses on criteria more than norms, on improvement more than ranking, and on collaboration more than competition. Indeed, when ongoing assessment is fully integrated into the teaching and learning process, it tends to shift the usual roles and responsibilities of teachers and learners significantly.

Teachers who wish to practice ongoing assessment as part of teaching for understanding must develop assessment tools and processes that truly support their understanding goals. For many teachers, devising assessment rubrics causes them to inspect and often revise their goals. As teachers define more explicitly the qualities they hope to see in their students' work, they sometimes identify tacit goals that they value but had not stated explicitly. Becoming more explicit about goals and assessment criteria is demanding intellectually. Dialogue with colleagues who

American Studies: The Harlem Renaissance[1]

How can students come to understand what it means to engage in critical thinking? This is one of the essential questions for high school students in this American Studies unit about the Harlem Renaissance as a social movement. Through Internet research, exploration of multimedia artifacts, on-line journaling, discussion, and the process of writing essays, students develop the ability to think critically about the social, historical, and cultural contexts that generated the far-reaching political and artistic contributions of African Americans in New York City during the Harlem Renaissance of the 1920s.

Throughout this unit, students use Web resources on the Harlem Renaissance with on-line tools (Filamentality and Nicenet) to support their study. Learning with these on-line technologies provides students ample opportunity to engage in self- and peer assessment. Ongoing assessments support students' development of critical perspectives, analytical skills, and the ability to form opinions and negotiate shared meaning with others.

Students begin their investigation with a series of informal and guided activities to familiarize them with historical events, important figures, and political thought, as well as creative works of architecture, poetry, jazz, drama, art, and literature related to the Harlem Renaissance. Their exploration is supported by an annotated "hotlist" of rich multimedia Internet resources developed by their teacher using the on-line tool Filamentality. The specific organization of the hotlist helps keep students focused on understanding goals and enables them to appreciate alternative perspectives for analyzing the many contributions of the Harlem Renaissance. Based on their research, students develop a series of questions in class discussion that critical thinkers would ask about the context, structure, meaning, and impact of artistic works, which their teacher uses as part of a rubric for assessing their essays.

As students read biographies, explore artistic forms, and engage in close examination of period works of their choice, they write public journal entries on an Internet site called Nicenet—a free on-line service that supports conferencing, sharing links and documents, and conducting threaded discussions. In the Nicenet forum, students comment on one another's journal entries, post reflective assessments of their own process of understanding, and read annotations posted by their teacher to guide their work. As a culminating performance, students write metacognitive essays analyzing the critical thinking of Harlem Renaissance figures, as expressed in their works, and reflecting on the students' own perspectives and critical thinking about the Harlem Renaissance.

Resources

Nicenet
http://www.nicenet.org/index.cfm

Harlem Renaissance
http://www.nku.edu/~diesmanj/harlem.html

Harlem Renaissance Resources
http://www.usc.edu/isd/archives/ethnicstudies/harlem.html

Filamentality
http://www.kn.pacbell.com/wired/fil/

---

[1]This vignette is based on the work of Miranda Whitmore and Janet Jehle.

are committed to similar practices is helpful, both as a means of formulating useful ideas about goals and assessment criteria and for learning about how these approaches lead to better educational results.

Sharing assessment criteria publicly is also challenging for many teachers. Some teachers feel that successful students ought to be able to intuit the qualities of good work without direct explanation or examples. Stating explicit assessment criteria evens the playing field somewhat by giving all students clear information about how to succeed, and it tends to promote better-quality work from all students. Some teachers may worry that being too explicit about goals and assessment criteria prevents teachers from evolving their plans in response to new insights that the class may generate as they pursue learning. There is no need to cast goals or assessment criteria in concrete, however; they can be modified as both teacher and students become more knowledgeable and aware of the main purpose of a lesson or curriculum unit.

Ongoing assessment gives students more responsibility for assessing and promoting their own learning and that of their fellow learners. Students may even contribute to defining assessment criteria. Developing the culture of a classroom to encourage this kind of sharing of intellectual expertise and responsibility may require a significant change,[2] especially in settings where teachers or students have not typically been expected to think for themselves or to exercise judgment.

When Kristi's students prepared work for authentic audiences, they were not simply working for a grade or obeying the teacher but rather producing work that they hoped would make a significant contribution in the real world.

Connecting the classroom with the outside world can be a way to help cultivate this kind of responsibility. When Kristi's students prepared work for authentic audiences, they were not simply working for a grade or obeying the teacher but rather producing work that they hoped would make a significant contribution in the real world. They were inspired to assess and improve their work because they wanted to be effective. Kristi also invited parents and other volunteers into her class to supplement the ongoing assessment that she herself could provide. When assessment rubrics are clear and public, outside helpers understand how to provide constructive suggestions for improving students' work. The participation of other helpful adults in the classroom is valuable for the time-intensive student consultations that are an important part of ongoing assessment.

Kristi modeled effective feedback strategies in her conferences with students. She intentionally taught them a step-wise process for interacting effectively with peers to improve their work, including how to do so with respect and positive support. Practicing peer assessment in reciprocal roles, as both a receiver and provider of constructive assessments, helps learners build their capacity to offer and benefit from ongoing assessment.

Questions for Reflection

1. How might you incorporate ongoing assessments that use clear, public criteria directly related to understanding goals, that engage learners themselves in peer and self-assessments, as well as teachers and other experts, and that generate constructive feedback to inform next steps in both students' and teachers' work?

2. How might new technologies support the processes of making student work visible for assessment, feedback, and revision?

NOTES

1. Dweck, C. S. *Self-Theories: Their Role in Motivation, Personality, and Development.* Philadelphia: Psychology Press, 1999.

2. Engaging students as thoughtful critics of their own work requires developing their confidence in their own intellectual judgment. Involving them in peer feedback depends on students' appreciation of the value of contributing to a community of learning rather than competing with fellow students. These beliefs may be very different from the usual culture of a school, in which case, collaborative ongoing assessment will require deliberately fostering a shift in values. For more discussion of these ideas, see "How Teaching for Understanding Changes the Rules in the Classroom" (Wiske, M. S. [ed.]. *Education Leadership,* 1994, *51*(5), 19–21).

Learning in Reflective, Collaborative Communities with New Technologies

Most learning takes place through reflective dialogue and collaboration, ideally in a community where trust, respect, and reciprocity are shared norms. Yet classrooms are places where talking with the outside world is usually impossible, and even talking with classmates may be discouraged. New communication technologies, including the Web, offer valuable means of overcoming the isolation of teachers and students in schools. Using new technologies to support reflective, collaborative communities is the focus of this chapter.

Learning is a social process that is mediated by language and advanced through interpreting and negotiating meaning with other people.[1] People learn by reflecting on what they know, considering ideas from multiple perspectives, and analyzing their experience with alternative interpretive frameworks. Collaborating with others enriches one's capacity to develop and apply ideas. Fellow learners may contribute different kinds of knowledge and expertise to a joint project and help one another clarify partially understood questions and insights. Collaborative learning is particularly valuable when it takes place within a community of learners who share common concerns and feel bound by shared norms and commitments, even if the members of the community may differ dramatically in their opinions and backgrounds.[2]

Engaging students in reflective, collaborative communities of learners is important, not only because it promotes their understanding of academic content but because such experiences also help students learn how to cooperate in teams to solve

TIP

Learning is a social process that is mediated by language and advanced through interpreting and negotiating meaning with other people.

TIP

Negotiating meaning across cultural differences often illuminates taken-for-granted assumptions.

problems and create products no one of them could accomplish alone. In their analysis of the "new basic skills" that employers want in their workers, Murnane and Levy[3] found that knowing how to communicate and collaborate effectively in groups was a high priority. Learning through exchange with other members of a community about issues that are important to the group helps students connect their academic work to authentic concerns, take advantage of a broad range of resources, and potentially make a significant contribution to their community. These advantages hold true, whether the community is simply other classmates or is based in the school, the local community, a national group, or an international network. When the learning community encompasses diverse languages, cultures, and locations, many additional benefits are possible. Negotiating meaning across cultural differences often illuminates taken-for-granted assumptions, demonstrates how varied and multifaceted "truth" may be, and helps students appreciate the similarities and differences of people around the world.

Teachers also benefit from participating in reflective, collaborative communities. Helping twenty to thirty, or more, varied students learn effectively is difficult for any lone teacher to accomplish. Drawing in other people—parents, experts, professional colleagues—from the school, the local community, and distant locations supplements the expertise and attention that the teacher has to offer. Finally, teachers' own learning advances through their engagement in reflective and collaborative professional communities. Such relationships help teachers examine and improve their own practices while exchanging ideas and resources with other thoughtful colleagues.[4]

Networked technologies provide multiple advantages for connecting learners with reflective, collaborative communities. Multimedia tools allow learners to express their ideas in a range of forms: drawings or diagrams, flowcharts or graphs, text formatted to highlight key ideas, or sounds, video, and hypertext with links that suggest alternative paths through a product.

TIP

Digital media enable students working together on a product to combine varied forms of expression, to take advantage of their different expressive strengths, and to create a product that communicates through varied media.

Nonverbal digital products can be especially helpful when students are communicating across language barriers. Digital media enable students working together on a product to combine varied forms of expression, to take advantage of their different expressive strengths, and to create a product that communicates through varied media. As they discuss ways that different representations serve to communicate about a concept or phenomenon, students may reflect on the content of their work as well as the process of understanding and communicating about it. Digital technologies capture work in forms that can be easily modified so that collaborating partners are able to change, combine, and rearrange components of their work until they are all satisfied with the final product.

Networked technologies support collaborative learning by providing access to information and resources for teachers and students, mediating communication,

Getting a MUVE On: Exploring Multicultural Perspectives Through Art[1]

Cultural and global awareness are among the literacies essential to success in the Digital Age, as described in the North Central Regional Education Laboratory's enGauge Web site and Framework for 21st Century Skills. Youngsters in the midst of developing their own identities may have difficulty appreciating experiences and cultures different from their own. In this unit, middle school students in the United States and Italy come to recognize one another's unique experiences in a virtual world called a MUVE (Multi-User Virtual Environment).

Students learn about art appreciation and criticism together in an interactive on-line universe they collaboratively design and create using the ActiveWorld's MUVE platform. Because art transcends language boundaries and also communicates the unique, culturally influenced creative vision of the artist, it is an ideal medium for students to express and discuss different perspectives across language and geographical divides. As students critique works of art, write compare-and-contrast essays, and exchange their own artworks with international peers, they collaboratively learn cultural awareness, communicate clearly across languages and cultures, and come to understand themselves and one another's experiences and points of view.

Students in both countries study Van Gogh's 1888 painting of his bedroom in Arles. They brainstorm and share their ideas about what Van Gogh's room and his painting communicate about his identity, culture, and the time and place where he lived by posting comments in a "room" in their virtual world. Using chat and e-mail, the students collaboratively generate a list of characteristics that contribute to people's identity. With this list of criteria in mind, students create their own bedroom self-portraits.

After a lesson by expert on-line mentors about how to use ActiveWorld's development tools, students design and assemble three-dimensional virtual representations of their own bedrooms, applying the group's criteria and selecting those items from among hundreds of "objects" that they believe best represent their cultural and individual identity. Students then tour each other's "rooms" and develop questions they pose to their international partners via synchronous chat and e-mail exchanges.

As a culminating performance, each student drafts, edits, and revises a compare-and-contrast essay about his or her virtual room, an international partner's virtual room, and Van Gogh's painting, incorporating standards for writing, analysis of the design process, identity criteria, and what they have learned about art and cultural perspectives. The final essays are posted in the virtual world. To celebrate and share their learning with others, students prepare and host a virtual open house to which they invite peers and parents from both nations into their new reflective, collaborative community for building cross-cultural understanding.

Resources

EnGauge Framework for 21st Century Literacies
http://www.ncrel.org/engauge/skills/skills.htm

ActiveWorld's Education Universe (AWEDU)
http://activeworlds.com/edu

Project Zero's Research Projects on Learning and the Arts

http://www.pz.harvard.edu/Research/ResearchArts.htm

National Gallery of Art Van Gogh Exhibition
http://www.nga.gov/exhibitions/vgbro.htm

Multicultural Literacy
http://www.kn.pacbell.com/wired/21stcent/cultural.html

---

[1]This vignette is based on the work of Amy Fritz, Beth-Ann Keane, Patricia Norris, and Kate Paterson.

and enabling learners to publish their work for varied audiences. The Internet offers access to resources that allow school children to focus their investigations on issues that are of real importance to members of their community. E-mail permits users to send and receive many-to-many messages and to do so quickly. Students can share ideas and work with many other students all over the world, exchanging multiple rounds of reflective dialogue. The Web, with digital images, video and audio recordings, and videoconferencing, also allows students and their teachers to publish and collaborate on work, opening up the possibility of communicating with a wide range of audiences outside the classroom.

Case Study: iEARN Friendship Through Education
Schools Outfitting Schools Partnership with Afghanistan

Learning through collaboration with teachers and students around the world was a regular feature in Kristi's teaching, in part because she wanted her students to understand how to be responsible global citizens but also because she believed this work directly supported all aspects of her curriculum. For these reasons, Kristi belonged to iEARN (International Education and Resource Network)—a program that links classes around the world in collaborative projects using networked technologies.

Through iEARN, Kristi learned about an initiative launched in the fall of 2002 by the U.S. government called Friendship Through Education. This program invited schools and young people in the United States and other countries around the world to learn from one another, using e-mail, Web sites, and videoconferencing to support communication. iEARN was asked to provide leadership for educational organizations in the Friendship Through Education (FtE) consortium.

Kristi and her students examined the FtE Web site (http://www.friendship througheducation.org/) to see how they might become involved. One of the FtE projects was Schools Outfitting Schools (SOS), accessed at http://www.friendship througheducation.org/sos.html, through which schools in the United States could raise funds to provide supplies for schools in Afghanistan. They learned that many children in Afghanistan were just beginning to return to school for the first time in years. There was a critical need for educational supplies in schools that were damaged and crumbling from years of neglect and war.

Kristi knew that her students could develop understanding of a whole range of goals in literacy, mathematics, and social studies through this project, while building the students' capacity to use a range of new technologies effectively. At the same time, the project would engage them in service learning and global citizenship that would both support their core academic learning and address one of her most heartfelt overarching goals for her students: understanding how to make a positive contribution in the world where the need for help was significant. The children decided to raise funds in their own class and try to engage the whole school and the wider community in the SOS project as well.

Kristi talked with her students to define learning goals they would pursue through this project. Their goals were to understand the following:

1. How to accurately read and interpret the implications of information in order to take appropriate action in responding to a global community need

2. How to use writing skills and new technologies to communicate effectively the importance of responding with action to a global community need, to document the process of meeting that need so that others could learn how to do similar projects, and to build ongoing collaboration with schools from the local community to schools in Afghanistan

3. How to calculate not only the costs of school supplies but the economic values of those supplies in Afghanistan and the educational value of providing those supplies

4. How to appreciate the importance and value of learning and making a difference with a collaborative local-to-global community of school peers, teachers, principals, and parents, as well as local, national, and international organizations involved with the SOS project (including UNICEF, iEARN, FtE, and Afghans for Civil Society)

With these goals in mind, Kristi and her students mapped out a sequence of steps and thought about how to get started on accomplishing their goals. They planned to involve the whole school and the community in raising funds. They would develop a Web site to support the project and a video to document their work throughout the year. New technologies would support the full range of activities that they planned.

Kristi's students began by meeting with the school principal and the student council to plan ways of involving the whole school and connecting with the community.

Kristi's students began by meeting with the school principal and the student council to plan ways of involving the whole school and connecting with the community. They used e-mail to work with their schoolmates on planning fundraising activities and to communicate with other participants at the national and global level involved in the FtE initiative and iEARN. Kristi's students used word processors to draft and revise speeches to present at the school assembly and to community organizations. Students from the local university came into the classroom to help Kristi's students develop artwork and prepare their speeches. Fifth-graders used computers to prepare and print flyers for the whole student body to take home. Teachers and students worked together developing a Web site to share information about the project. They made digital images and video of multiple steps in the project to include on the Web site and to produce a video about the project.

Monetary contributions began to flow. Kristi's students counted the coins and used a computer to keep a running tally of their funds. As the SOS funds provided school supplies for children and teachers in Afghanistan, the staff of an organization called Afghans for Civil Society (http://www.iearn.org/afghan/ iEARNAfghanistan.html) e-mailed frequent reports about these activities. This organization also sent Kristi's class digital images and video of children learning in their rebuilt schools in Afghanistan, many of which were opening for the first time in years. These messages and images were added to the Web site documenting the project.

For Kristi's students, development of the project Web site became an opportunity to capture, analyze, and share what they were learning through this project (http://www.psd267.wednet.edu/%7Ekfranz/Schoolyear0102/socialstudies0102/ SOS/SOS.html). Teachers and students worked together on authoring text and assembling digital images to document the urgent need and the students' valuable response to this need. Students practiced both reading and writing skills as they developed, read, discussed, and revised Web pages about the project. At the same time, they learned about Web design and how to use the Web to communicate clearly and effectively.

This project developed multiple layers of reflective collaborative communities: classroom, school, local community, as well as national community coordinated through the U.S. Department of Education and international community with support from iEARN, Afghans for Civil Society, and others. Within and across these levels, new technologies supported reflection, communication in multiple media, and exchange of ideas, resources, and information. As a result of this project, a team of educators, including participants from iEARN, Afghans for Civil Society, and the Harvard University Center for Middle Eastern Studies, joined together to develop a curriculum for cross-cultural school partnerships to support ongoing collaborative learning involving schools in the United States and Afghanistan.

Key Features of Reflective, Collaborative Communities

The key features of reflective, collaborative communities of learners are clearly suggested by the terms themselves. *Reflection* is a process of standing back from experience and examining it in ways that generate meaningful knowledge. Usually, reflection requires representing experience in some way, often with language but perhaps nonlinguistically, with images or video, or in music or dance, or with gestures. Representing experience entails selecting aspects of experience, relating them, and expressing them in some form that communicates one's memory and interpretation of experience. Through such a process, one comes to crystallize ideas and connect them to other ideas and prior experiences. Reflection is a way of making sense, beyond merely experiencing or acting intuitively. It connects "hands-on" activity with "minds-on" thinking to form knowledge that can be extracted from the concrete here-and-now and packaged in a more generalizable, transferable form. For learners, reflection is an important way of formulating evolving ideas in ways that can be inspected, analyzed, discussed, and revised to develop meaningful understanding within a community. Kristi's students reflected on their understanding of reading and writing, mathematics, and the meaning of community through representing their ideas in multiple ways, for example, e-mail, Web sites, and oral presentations.

Reflection can be accomplished alone, but its contribution to learning is often enhanced by working with others on the process. *Collaboration* involves interactions with other people, reciprocal exchanges of support and ideas, joint work on the development of performances and products, and co-construction of understanding through comparing alternative ideas, interpretations, and representations. Peers help one another express their ideas, compare alternative interpretations, and learn from seeing multiple approaches to products and performances. Kristi's students often collaborated with other students—both classmates and peers in other places—so that they learned through dialogue about their shared work.

Collaboration is most satisfying and educational when it takes place within a *community* rather than in a loose assemblage of separate individuals. Communities are groups of people who recognize and appreciate both their diversity and their commonalities, who share a commitment to one another's well-being, and who trust one another to use and develop the communities' resources in ways that benefit the group. Members of communities typically share a common language (or some means of communication) and endorse some common values.

Kristi carefully built a community within her classroom through modeling and teaching the values, attitudes, and skills that enabled her students to work and learn together. She systematically engaged her class in forming connections with multiple communities: in the school, in the town, and in the global community. She connected every major learning project to issues that were important to one or more of these communities. In this way, her students' learning became meaningful

---

Key Features of Reflective, Collaborative Communities

- *Reflection:* Process of standing back from experience and examining it in ways that generate meaningful knowledge
- *Collaboration:* Co-construction of understanding through comparing and combining alternative ideas, interpretations, and representations
- *Community:* Groups of people who recognize and appreciate both their diversity and their commonalities

The Oral History Project[1]

The work that historians do to study how the past shapes values, beliefs, and behavior today can seem abstract to students. This social studies unit involves eighth-graders in creating authentic oral histories in order to appreciate how family stories connect their lives to history while strengthening their connection to the elders in their community. Students also come to understand that oral narratives are historic artifacts; they learn how they differ from written accounts and how historians gather, preserve, and interpret primary sources.

The Oral History project is taught by an interdisciplinary team, including language arts and history teachers, the school librarian, and special education and ESL instructors. The project is organized as a WebQuest—a series of Web pages with instructions, questions, exercises, assessments, tutorials, and electronic and print resources that students access and revisit throughout the project. The WebQuest makes the knowledge, methods, forms, and purposes of the unit explicit and scaffolds students' work for a sequence of tasks.

Students begin by reading diverse folktales and participating in an interactive performance and discussion with a storyteller, who demonstrates the impact and role of stories in history. Then students study ancient civilizations, working in groups to examine "cultural universals," such as family relations, politics, economics, geography, and arts and food; they look for patterns across civilizations and historic periods.

Next, students prepare to conduct oral histories, first exploring models on the Internet. They watch videotaped examples their teachers have prepared of effective and ineffective interviews, then work in small groups to draft, critique, and revise interview questions and to practice interviewing skills. They learn to use digital video and still cameras, multimedia software, voice recorders, and scanners to record and edit their oral histories. Each student interviews an older family member, researches the relative's cultural and historic context, and examines artifacts from the interviewee's life and times. Students select meaningful quotes and narrative elements, then construct multimedia presentations using iMovie®, QuickTime®, and PowerPoint® that incorporate selections from the narratives and background information with their own reflections and interpretations.

The unit's culminating event is a community open house where students present their oral history productions. Students and their guests enjoy family recipes and celebrate the class's work. Finally, students articulate their connection across generations in letters to their interviewees reflecting on what they learned from their life stories and the process of creating oral histories.

Resources

Myths, Folktales, and Fairy Tales
http://teacher.scholastic.com/writewit/mff/

Historically Speaking WebQuest http://wms.watertown.k12.ma.us/library/oralhistory/index.html

Oral History Association
http://omega.dickinson.edu/organizations/oha/

Library of Congress American Memory Collection Learning Page
http://memory.loc.gov/ammem/ndlpedu/index.html

Making Sense of Oral Histories
http://historymatters.gmu.edu/mse/oral/

An Oral History Primer
http://library.ucsc.edu/reg-hist/ohprimer.html

Oral History in the Teaching of US History
http://www.ericfacility.net/databases/ERIC_Digests/ed393781.html

Great American Speeches Archive
http://www.pbs.org/greatspeeches/timeline/

Smithsonian American History Timeline (images and artifacts) http://www.si.edu/resource/faq/nmah/timeline.htm	Songs in American History http://history.sandiego.edu/gen/snd/a-songlist.html
Facing History and Ourselves http://www.facing.org/	WebQuest Page http://webquest.sdsu.edu

[1]This vignette is based on the work of Katie Beller, Ellen Fitanidies, Allison Levit, and Linda Picceri.

because it was anchored in real concerns that mattered to the children. Their learning was also enriched by bringing resources from the community into the classroom, either literally or virtually.

Kristi's students began by meeting with the school principal and the student council to plan ways of involving the whole school and connecting with the community. In order to design and support learning of these kinds, teachers themselves need to be part of reflective, collaborative communities of professionals who share their values. One reason that Kristi participated in iEARN was so that she could develop a supportive professional community while connecting her students with a global community of peers. Further discussion of this issue appears in Chapter Eight.

How New Technologies Support Reflective, Collaborative Communities

Tools both shape and enable the way we think and communicate. New technologies alter the ways we access, organize, analyze, present, share, and transform knowledge. Understanding how to make use of networked technologies to foster reflective and collaborative learning is essential for success, satisfaction, and responsible citizenship in a global society.

E-mail and on-line threaded discussion tools create a new kind of communication forum. Communication in these environments takes place via written messages that are more deliberate than spoken language but less formal than most written texts. For this reason, e-mail encourages the expression of "first draft" thinking. Composing e-mail may draw out writing from learners who do not feel confident about producing work that must meet a higher standard of completeness and correctness. Yet because on-line messages are written, they can be crafted with more thought and reflective care than spoken language. Thus networked technologies may promote a more analytical dialogue than face-to-face oral conversations.

On-line communication is automatically captured and archived in forms that can be reviewed and revised. This enables learners to look back at a sequence of

TIP

Learners who are creators,
as well as consumers, of
information on the Web have
rich opportunities to develop
a sense of community with
others outside their classroom.

messages and reflect on their meaning. By facilitating the exchange of multimedia materials, networked technologies facilitate learners' abilities to view multiple perspectives and interpretations. Digital technologies also capture work in forms that can be easily combined and revised, thereby facilitating collaboration.

Kristi's students developed intense relationships with students in other parts of the world through exchanging pictures, video, and artwork, as well as written messages. Photographs of the schools and children in Afghanistan, along with reports about their conditions, helped Kristi's students feel connected to their Afghani counterparts. The immediacy of exchanging messages with distant peers also contributed to a feeling of community.

Learners who are creators, as well as consumers, of information on the Web have rich opportunities to develop a sense of community with others outside their classroom. Kristi's students combined pictures of artwork, photographs, video, and written text to create a Web site about their SOS project. Constructing and linking Web pages helped the students compare, interpret, and convey their varied perspectives on the project. Producing video and building the Web site also enabled the students to communicate about this project to others who might add their support to the cause, such as local charitable organizations.

Overall, the advent of digital multimedia and the Internet provide unprecedented opportunities for learners to develop understanding through representing ideas with multiple kinds of artifacts, archiving messages, communicating via both synchronous and asynchronous dialogue with large numbers of people all over the globe, and publishing their work so that it can make an impact in the world.

Planning and Teaching in Reflective, Collaborative Communities with New Technologies

Kristi's project with iEARN's Friendship Through Education involved her students in work that differed dramatically from what goes on in most school classrooms. Indeed, teachers must address a range of issues in order to take full advantage of new technologies to promote learning through reflective, collaborative communities. Of course, access to networked technologies and to the technical assistance to keep them working smoothly are important. But there are additional, less obvious issues that are equally essential and sometimes even more difficult to address than technical resources.

Technical Resources

Networked technologies require the support of a community. The hardware and the means for gaining access to Internet service providers change continually, and they are often complicated to work with. "You really need a community in order to learn how to work with new technologies and to be able to use them effectively in a school context," says Kristi. Teachers may need local technical assistance staff to help configure their computers and to provide regular updates about new advances, for example, with wireless networks that use radio waves to connect to hard-wired communication networks or to satellites. Kristi also relied on parents and friends to help her learn how to use networked technologies, for example, to create a Web page, to take advantage of the synchronous communication facilities, and to participate in videoconferences first using CUSeeMe and later using the Washington State K–20 Videoconferencing Network. The point is that these technologies are continually evolving. Educators must build relationships with others who are tracking technological developments and rely on such colleagues to help them acquire and learn what they need in order to use new technologies to serve their own educational agenda.

Time for Planning

If teachers are going to work effectively with other teachers, they need time to think together about their curricular goals and how they will help their students achieve these goals. Some planning can be accomplished through asynchronous collaboration with colleagues, perhaps using networked technologies to exchange ideas and store materials. But some real-time dialogue is usually necessary as well. In many schools, even neighboring teachers can rarely find time to meet, because their schedules are full and their few free blocks of time do not coincide. Kristi worked with the schedule planner in her school to try to arrange common free periods with her colleagues. Sometimes this was accomplished by sending their classes to different specialists—art, physical education, or music—during the same period. The teachers' union sometimes made the provision of some common planning periods a priority during their negotiation of contracts. In some cases, a special grant provided funding that allowed Kristi to hire a co-teacher part-time so that Kristi could meet with other teachers and develop curriculum materials for collaborative projects.

The usual class period of approximately forty-five minutes is not long enough to support extensive planning and collaboration. Some states and districts pay teachers for one or more days per year to participate in joint planning or professional development. This kind of arrangement sends a strong signal that teachers' efforts to extend their own learning and to provide innovative learning experiences for students are valued by local taxpayers.

The usual class period of approximately forty-five minutes is not long enough to support extensive planning and collaboration.

Changing Norms of Isolation and Competition

There are several reasons why collaboration and dialogue are not more common in schools. Some of the reasons have to do with access to resources—technology, time, and funding—that enable easy communication. Effective dialogue between

TIP

Networked technologies require the support of a community. The hardware and the means for gaining access to Internet service providers change continually, and they are often complicated to work with.

TIP

If teachers are going to work effectively with other teachers, they need time to think together about their curricular goals and how they will help their students achieve these goals.

students in classrooms and others outside the school does require access to such resources, to be sure. But even when these elements exist, school classrooms are often places where isolation and competition discourage students from discussing their work with classmates or collaborating on their work. This pattern reflects deep-seated beliefs and cultural norms about the nature of knowledge, the process of learning, and the means of gauging students' achievements.

Many schools reflect beliefs that knowledge is a private commodity, teaching is telling, learning is a race to the right answer, and achievement is measured by individual scores on competitive tests. These beliefs may not be explicitly stated, but if one examines the way administrators and teachers behave, the rules they set in classrooms, and the basis on which students are assessed, one is likely to conclude that these tacit beliefs are consistent with the culture in many schools. In such a culture, not knowing the right answer is embarrassing. Students feel humiliated if they can't shoot up their hands fast when the teacher poses a question. Worse yet is to be called on and say the "wrong" answer or "I don't know." Sharing "first draft" ideas that may be vague, incomplete, or shallow is also risky in an atmosphere that values only polished and correct work. If students are ashamed about not knowing the right answer, teachers are even less willing to display or acknowledge their ignorance in a culture where knowledge is a private commodity and a cause for high status.

Such a culture works against collaborative learning and free-flowing dialogue to co-develop understanding. It prevents students from participating in the very kinds of interactions and reflective conversation that are educationally beneficial to them. It may seal students off from the questions, beginning ideas, and creative contributions of their fellow learners. And it tends to block out the wealth of resources, mentors, colleagues, and authentic audiences in the outside world that could make schoolwork more interesting, effective, and valuable.

In order to overcome norms of competition and isolation, teachers must embrace and cultivate norms that honor the contributions of all members of the learning community, including every student in the class, as well as valuable collaborators outside the class. Teachers can begin this process by listening carefully to their students and finding ways to build on the ideas and other resources that students bring to the learning experience. In addition to modeling such collaborative modes of interaction, teachers may need to teach students explicitly how to work collaboratively, for example, by dividing and sharing roles, treating one another with respect, and taking turns.[5]

Teachers who want to foster collaborative learning and reflective dialogue will also have to confront the pressure in schools to focus primarily on raising students' scores on norm-based assessment tests. Such tests measure individual achievement and often pit students, classes, teachers, schools, and districts against one another in a competitive contest. Individual grades and ratings may motivate some students and teachers to be accountable for their own achievement. However, competitive ratings of these kinds may also undermine the cooperation and communal learning advocated in this chapter. Teachers who work in schools or districts where competitive assessments are prominent may need to search for ways to demonstrate how collaborative work ultimately helps all students achieve at higher levels while attempting to de-emphasize competitive rankings as much as possible.

Questions for Reflection

1. How might your learners benefit from opportunities to reflect, collaborate, and communicate with other people as a way to increase their understanding of priority goals?

2. How could new technologies support these forms of thoughtful interaction?

NOTES

1. There is a long history of stimulating analysis about learning in relation to social interaction and communication and mediating tools, including symbolic tools such as language and concrete tools like books or other information technologies. Important authors include the following three: (1) John Dewey (*The Child and the Curriculum, and the School and Society.* Chicago: University of Chicago Press, 1956; *How We Think: A Restatement of the Relation of Reflective Thinking to the Educative Process.* Boston: Heath, 1933); (2) L. S. Vygotsky (*Thought and Language* [trans., revised, and edited by A. Kozulin]. Cambridge, Mass.: MIT Press, 1986), and (3) Yrgo Engestrom (*Perspectives on Activity Theory.* Engestrom, Y., Miettinen, R., and Punamaki, R.-L. [eds.]. Cambridge, Mass.: Cambridge University Press, 1999).

2. For studies of the way people learn through participating in collaboration in communities of practice, see *Situated Learning: Legitimate Peripheral Participation* (Lave, J., and Wenger, E. Cambridge, Mass.: Cambridge University Press, 1991) and *Communities of Practice: Learning, Meaning, and Identity* (Wenger, E. Cambridge: Cambridge University Press, 1998).

3. Murnane, R. J., and Levy, F. *Teaching the New Basic Skills: Principles for Educating Children to Thrive in a Changing Economy.* New York: Martin Kessler Books, Free Press, 1996.

4. Schön, D. A. *The Reflective Practitioner: How Professionals Think in Action.* New York: Basic Books, 1983.

5. Brown, A. L. "Transforming Schools into Communities of Thinking and Learning about Serious Matters." *American Psychologist,* 1997, *52*(4), 399–413.

THREE

Learning to Teach for Understanding

How Teachers Learn to Teach with New Technologies

All educators who read this book will recognize some of the ideas and examples as similar to their own practices. Perhaps they will also see some ways they might deepen their goals, refine curriculum designs to align the components and focus them more directly on understanding goals, and extend their ways of using new technologies to improve students' learning. Some readers may perceive that many kinds of teaching and learning advocated and illustrated on these pages are significantly different from their current teaching. They may even conclude that such approaches are impossible within their own situations, especially if their contexts emphasize teaching as didactic instruction of a required curriculum and learning as measured by students' scores on multiple-choice tests.

It is important to recognize that learning to teach for understanding with new technologies is a gradual process. It can begin with minor changes and evolve over time through trial-and-error and success.

Even in such circumstances, however, some aspects of the approach described in this book can be tried. It is important to recognize that learning to teach for understanding with new technologies is a gradual process. It can begin with minor changes and evolve over time through trial-and-error and success. Ambitious, polished practice requires a mixture of materials, technologies, various forms of assistance, and evolving understanding. Teachers integrate these components gradually through seeking help, risking change, reflecting on the results, and risking change again.

Helping teachers undertake ambitious, adventurous, and reflective teaching with new technologies is itself a process of teaching for understanding. It cannot

Tips on Learning to Teach for Understanding with New Technologies

- It is important to recognize that learning to teach for understanding with new technologies is a gradual process.
- Teachers integrate these components gradually through seeking help, risking change, reflecting on the results, and risking change again.
- Learning to teach in these ways depends on cycles of thinking about learning and teaching, analyzing key concepts and methods of inquiry within or across subject matters, trying out new practices, and analyzing those experiences with like-minded colleagues and coaches.

be accomplished through transmitting information in the typical one-shot workshop, or even a short series of workshops. It is not primarily a matter of technical training. Learning to teach in these ways depends on cycles of thinking about learning and teaching, analyzing key concepts and methods of inquiry within or across subject matters, trying out new practices, and analyzing those experiences with like-minded colleagues and coaches.

Such cycles of professional support are difficult to provide for a host of reasons: effective coaches are rare; teachers' lives are busy; time for professional development is scarce; and travel is expensive. Even if supportive professional development can be orchestrated for a small number of teachers, successfully scaling up such efforts is very difficult. The usual "cascade model" of scaling up by "training the trainers" runs the risk of diluting the essential components of professional development as later tiers in the cascade become remote from the origins of the approach.[1]

New technologies offer significant advantages in overcoming these barriers to professional development. In addition, teachers learn about how to teach with new technologies from working with these technologies as part of the professional development process. Consequently, helping teachers learn how to teach for understanding with new technologies mirrors the same processes that are advocated throughout this book. Explicitly designing professional development to model the approaches that teachers are expected to enact with their own students also has the advantage of allowing teachers to learn from reflecting on their own experience.

Professional Development as a Process of Teaching for Understanding

Applying the principles of the Teaching for Understanding framework to the design of learning experiences for teachers requires only a slight shift of focus from using this framework to help K–12 students understand academic subject matter. The

> ## Teaching for Understanding Is Effective for Teachers' Professional Development
>
> Helping teachers learn how to teach for understanding with new technologies mirrors the same processes that are advocated throughout this book:
>
> - Organizing teachers' learning around topics that are generative for the teachers
> - Focusing on explicit understanding goals for teachers
> - Providing many opportunities for teachers to apply what they are learning
> - Building in ongoing assessment for teachers with constructive feedback
> - Helping teachers learn from one another in collaborative reflective communities

focus of learning shifts from school subjects to effective teaching, but all the basic principles of teaching for understanding still apply. Organizing teachers' learning around topics that are generative for the teachers, focusing on explicit understanding goals, providing many opportunities for teachers to apply what they are learning, and building in ongoing assessment with constructive feedback are all effective elements of professional development. Helping teachers learn from one another in collaborative reflective communities is an especially valuable component of professional learning for educators.[2]

Generative Topics

Organizing learning around a generative topic requires clear links between the focus of study and the learners' own experience, concerns, priorities, and circumstances. Learning to teach for understanding with new technologies can be approached from multiple entry points, depending on the interests of the teachers. Some teachers may be worried about teaching to required curriculum standards or mandated student achievement tests; for them, "teaching to standards with new technology" may be a generative topic. Indeed, this is the title of a course Stone Wiske designed and taught through an on-line professional development program that is described later in this chapter. This title emphasizes that integrating new technologies need not be seen as yet another requirement, in competition with other mandates, but rather as an effective means for meeting these priorities.

For many teachers, the idea of understanding as a performance capability and the whole approach of teaching for understanding are not at all generative. Teachers may believe that good teaching involves delivering content to students and that achievement is demonstrated by students' ability to produce correct answers quickly. Teachers may embrace this focus on "instruction" by teachers rather than "construction" by learners for many reasons: most teachers experienced "instruction" from their own teachers; many common vocabulary terms and expectations around schooling reinforce images of teachers as transmitters of knowledge (such

TIP

Integrating new technologies need not be seen as yet another requirement, in competition with other mandates, but rather as an effective means for meeting these priorities.

as "covering the curriculum" and "delivering instruction"); many school systems treat teachers as conveyors of centrally determined policies rather than professional artisans, and relatively few of the test items that most students are required to take measure understanding as a capacity to apply knowledge flexibly. Furthermore, when one teacher faces twenty, thirty, or even more students—perhaps several different classes of this size every day—delivering content may seem much more manageable than attempting to foster each student's understanding.

Many school systems treat teachers as conveyors of centrally determined policies rather than professional artisans.

Considering all these forces, it is an impressive testament to teachers' dedication that so many of them do care about their students' capacity to make sense, to apply what they learn in school, and to be able to use their knowledge flexibly and creatively in a range of circumstances. One way to help teachers who seem stuck in "instruction" mode is to ask them to think about something they understand well. For example, a science teacher might understand Newton's laws of mechanics thoroughly. Other examples are how to interpret poetry, how to play a musical instrument, how to coach a sport or artistic performance, or how to engage shy children. Then ask the teachers how they know that they understand this well. If possible, invite them to talk about their answers to these questions with one or two others. When teachers share their answers, they often say things like, "I can explain it in different ways to people who don't understand" or "I can do this under a wide range of circumstances" or "Other people who watch me tell me I'm good at it." It is usually easy to help teachers notice that their evidence of understanding entails performances.

TIP

To reveal teachers' ideas about the process of developing understanding, ask, "How did you come to understand something that you understand very well?"

To reveal teachers' ideas about the process of developing understanding, ask, "How did you come to understand this?" Inevitably, many elements of the Teaching for Understanding framework appear in the answers to this question: "I found this fascinating and I really worked at it"; "I got wonderful help from a coach who watched me try and then gave me tips"; "I practiced with a group of people and we helped each other." From such answers, it is easy to point out the patterns of engaging learners in cycles of activities that entail applying knowledge and skills in the context of authentic performances with rounds of assessment and feedback from an experienced coach and peers.

Most teachers who engage in a conversation around these questions about the nature and process of understanding will agree that their ultimate goals for their own students include a desire to foster this kind of learning, that is, a capacity to apply their knowledge flexibly and creatively in a range of circumstances. Most teachers will also acknowledge that students cannot develop such understanding simply through receiving good instruction.

Teachers may feel frustrated that they cannot see how to focus on "understanding goals" in the midst of pressures to "cover the curriculum" or "prepare students for the test." If teachers acknowledge that they embrace understanding as a goal in their own hearts, however, the stage is set to make teaching for understanding a generative topic.

Understanding Goals

Defining the goals of professional development so that they focus on understanding requires attention to the performances that participants will develop. Articulating these goals up-front and making them public helps to focus all the participants on these aims (including the leaders and the learners). What will teachers be able to do as a result of the experience? Are these goals mere skills (such as knowing some core features of a word processing program), or are they important understandings (such as understanding how to teach with a word processor effectively to help students write more vividly and clearly)?

Understanding goals for teachers may encompass multiple dimensions, just as teachers' goals for their students do. Goals for participants in professional development may focus on understanding particular knowledge, for example, subject matter content, a pedagogical concept like generative curriculum topics, or criteria for assessing the merits of an educational technology in relation to one's own purposes. Or goals may focus on pedagogical methods, such as understanding how to design and implement some ongoing assessment activities in which students serve as assessors of their own work. Some goals may focus on helping teachers appreciate and be able to participate in reflective collaboration with colleagues as a continuing process of professional inquiry. Understanding how to use modern communication technologies to improve both professional practice and students' learning may also be target goals. Whatever the goals may be, defining them clearly and making sure they focus on important understandings are ways to ensure that

---

Understanding Goals for Teachers

Understanding goals for teachers may encompass multiple dimensions, just as teachers' goals for their students do. Goals for participants in professional development may focus on

- Understanding particular knowledge
- Learning effective teaching methods
- Appreciating and being able to participate in reflective collaboration with colleagues as a continuing process of professional inquiry
- Understanding how to use modern communication technologies to improve both professional practice and students' learning

professional development activities aim for worthy purposes, not just training in isolated skills.

Of course, goals for teachers may include learning some concrete skills, such as learning how to communicate effectively with e-mail or how to upload digital images to a computer and modify them with software. What is important in professional development for understanding is that such skills are clearly connected to larger understanding goals, rather than being ends in themselves. Teaching skills by helping teachers produce something that they will be able to use in their own context is often a good way to make learning generative while connecting the development of skills to larger goals of improving professional performance.

Making understanding goals public may seem awkward or overly "teacherly" when working with adults. In our experience, however, when the goals truly focus on understanding, stating them up-front tends to engage teachers by helping them see the relevance of the activity to their own practice. Eventually, the experience of being taught with clear understanding goals helps teachers appreciate the value of this element of the Teaching for Understanding framework.

One final word about explicit goals: it is not necessary to make learning about the Teaching for Understanding framework a front-and-center, explicit goal. In fact, drawing too much attention to the framework itself can be counterproductive if the primary purpose is to help teachers begin seeing ways of integrating technology or thinking more deeply about their subject matter. If the explicit focus is not on pedagogy, it may be more effective to apply Teaching for Understanding principles in designing and conducting professional development but not to make the understanding of these principles an explicit goal at the beginning. If a secondary goal is for teachers to notice how the professional development experience reflects the elements of Teaching for Understanding, it may be effective to draw attention to the framework later in the process, after teachers have begun to demonstrate understanding of the primary professional development goals.

Performances of Understanding

Professional development built around Teaching for Understanding with new technologies includes a sequence of activities that enable participants to apply what they are learning in their own context. Just as students may benefit from a series of performances that builds up from what they already know about the targeted understanding goals, professionals may benefit from early experiences based on their current practice and concerns. For example, teachers might first be asked to reflect on ways in which their current practices or their most pressing professional concerns do and do not relate to elements of the Teaching for Understanding framework. They might choose a curriculum unit they have already taught and analyze it in relation to the framework elements. Teachers could also learn about one of the framework elements and then be asked to apply it in their classroom and exchange reflections about how the process went. Any of these activities could serve as "introductory performances" that would help teachers build on their existing knowledge while beginning to learn about Teaching for Understanding. Such activities also give the leaders of professional development opportunities to learn how teachers currently think and talk about their practice so that subsequent activities can be designed to build on this foundation. Then teachers might learn about one or more technologies they could use to improve students' understanding. "Guided inquiry" performances might involve coaching teachers as they gradually redesign a curriculum unit, integrating a new technology so that the unit more fully meets the criteria for all elements of the Teaching for Understanding framework. As a "culminating performance" of understanding, participants might assess a fellow learner's curriculum design, using a rubric based on the elements of Teaching for Understanding, and provide feedback about the design's strengths and ways it might be improved. Of course, actually teaching the unit is an excellent culminating performance, especially if the process includes reflecting with colleagues on how the experience related to principles of Teaching for Understanding with new technologies.

Performances of Understanding for Teachers

Professional development built around Teaching for Understanding with new technologies includes a sequence of activities that enable participants to apply what they are learning in their own context:

Introductory Performances

- Choose a curriculum unit they have already taught and analyze it in relation to the Teaching for Understanding framework elements.
- Learn about one of the framework elements and then apply it in their classroom and exchange reflections about how the process went.

Guided Inquiry Performances

- Coach teachers as they gradually redesign a curriculum unit, integrating a new technology so that the unit more fully meets the criteria for all elements of the Teaching for Understanding framework.

Culminating Performances

- Assess a fellow learner's curriculum design, using a rubric based on the elements of Teaching for Understanding, and provide feedback about its strengths and ways it might be improved.
- Teach a unit designed with the Teaching for Understanding framework-designed unit in their classroom and reflect on the experience.

Ongoing Assessment

Using explicit criteria to assess their own performance and participating in peer assessments are not customary experiences for teachers. Although in some schools teachers assess their practice collaboratively, many teachers still live surrounded by "norms of privacy," as Judith Warren Little[3] calls them, which discourage talking about teaching practices and prevent teachers from learning through dialogue with fellow professionals. Educators may also feel that teachers are "born not made," that teaching is more an art than a science, and that great teaching practice cannot be described with explicit criteria.

Teachers, like their students, need to be helped to feel safe in sharing their questions and doubts with peers.

In order to help teachers learn from ongoing assessment of their performances, such beliefs must be confronted. Teachers, like their students, need to be helped to feel safe in sharing their questions and doubts with peers. Professional development leaders can help to cultivate this kind of trust by acknowledging their own incomplete knowledge and noting when they have learned from their own students. Teachers also need to develop specific vocabulary and norms for talking in detail about their practice, in order to be able to help one another devise better ways of teaching.[4] Developing a list of the qualities of effective designs for learning and using it to analyze curriculum plans is a valuable way to help teachers improve. Listing the qualities of generative topics, for example, and then using the list to analyze a curriculum unit help teachers understand the meaning of those qualities and develop their capacity to reflect on their practice in relation to explicit criteria.

Ongoing assessment in a professional community depends on effective coaching by experienced mentors. Peer feedback with colleagues who are at similar levels of understanding can also be particularly effective, just as it is with younger learners.

TIP

Developing a list of the qualities of effective designs for learning and using it to analyze curriculum plans is a valuable way to help teachers improve.

Reflective, Collaborative Communities

Professional development that involves teachers in reflection and collaboration with colleagues helps them dispel the norms of privacy, understand from one another how to apply a more analytical approach to their own teaching practice, and begin to appreciate the benefits of collegial exchange. The elements of the Teaching for Understanding framework provide a professional community with a particular language that enables them to discuss aspects of curriculum design and teaching activities in focused and specific ways. Although terms like *performances of understanding* may seem clumsy at first, they come to signify a specific kind of learning activity. As teachers become more familiar with the criteria for these performances, the term takes on practical meaning. It no longer seems so awkward, and it helps teachers realize that they are aiming for a particular type of learning activity with their students—

one that is minds-on, not just hands-on work. As teachers work with the Teaching for Understanding framework, they usually revert to abbreviations like TfU, UGs (for understanding goals), UPs (for understanding performances or performances of understanding), and OA (for ongoing assessment). That's a signal that the terms have come to serve as a useful language for specific, focused conversations about how to define goals, design teaching, and conduct meaningful assessments.

Case Study: On-Line Professional Development Resources for Teachers

The WIDE (Wide-scale Interactive Development for Educators) World project at the Harvard Graduate School of Education offers professional development for educators that both models and teaches the Teaching for Understanding framework using new technologies. WIDE World began in 1999 as an effort to connect educational research on teaching and learning to the improvement of practice. Its founders, Stone Wiske and David Perkins, had long been involved in such efforts; since the mid-1990s, they had explored ways of using the Internet to fortify bridges between educational research and practice. Stone initiated a Web site called ENT (Education with New Technologies at http://learnweb.harvard.edu/ent) to support educators interested in teaching for understanding with new technologies. Both the ENT on-line learning environment and WIDE World on-line professional development courses were deliberately designed to apply Teaching for Understanding and to use Web-based technologies to support professional development. These two projects illustrate some of the rich variety of ways that new technologies can be applied to make professional development more effectively enact principles of teaching for understanding.

Education with New Technologies Web Site

The ENT on-line learning environment includes a range of resources, interactive tools, multimedia case studies, and electronic forums that are freely available to anyone with an Internet connection. The site explicitly models the principles of the Teaching for Understanding framework to help educators learn about it as a structure for guiding the integration of new technologies.

Components of the ENT site are organized within the context of a village metaphor, with buildings for getting oriented, reviewing resources, meeting, and learning. The village metaphor was chosen to be immediately engaging and accessible so that even teachers with very little experience using new technologies would find the site approachable and generative.

The designers of the ENT Village had several understanding goals in mind. They hoped users who registered in the site would understand how to use the Teaching for Understanding framework to develop curriculum that integrates new technologies specifically to improve students' learning. They also hoped educators would understand how to collaborate with like-minded colleagues in the process of ongoing curriculum design and analysis. Several components of the site directly support these goals.

Images of possibilities help educators begin to envision ways that the principles of Teaching for Understanding and new technologies might be applied to their own subject matter, learners, and contexts.

For example, the Gallery in the ENT Village contains multimedia pictures of practice illustrating a range of ways that educators use the Teaching for Understanding framework to structure their work with new technologies. Links embedded in these case studies provide more details about lesson plans and materials, insights into the teacher's experience and rationale, and examples of students' perceptions and products. These images of possibilities help educators begin to envision ways that the principles of Teaching for Understanding and new technologies might be applied to their own subject matter, learners, and contexts.

An important tool for helping educators learn how to apply the framework's original four elements to their own practice is the Collaborative Curriculum Design Tool (CCDT), found in the ENT Workshop. This interactive on-line tool provides a workspace where users can design curriculum around each element of the framework: generative curriculum topics, understanding goals, performances of understanding, and ongoing assessment. The tool provides several forms of support and assistance that users can call up if they wish, such as examples, prompts, and criteria for each element. The structure and resources built into the tool help educators keep the goals of Teaching for Understanding in mind as they design curriculum. The CCDT also provides links to curriculum standards in different subject matters from various educational organizations, as well as resources for identifying appropriate educational technologies. These components help teachers connect their designs to their own local curriculum requirements.

To foster collaboration and reflection with fellow professionals, the CCDT incorporates a built-in threaded message board so that users can communicate easily with other members whom the initiator of the unit chooses to add to a design team. For example, the designer might choose to add a coach, fellow learners, and other collaborators. Through the message board, members of the design team can record and catalogue their questions, feedback, and suggestions about an emerging curriculum design. Users may also attach links and files to their design. Because the tool is on-line, members of a design team can assemble all their materials in one place and communicate freely with one another, even if they are geographically dispersed and unable to meet simultaneously. A recent version of the CCDT is structured to help users refine and align elements of their curriculum to be sure that goals, activities, and assessments all work together coherently. Overall, the CCDT is one example of a way to use on-line resources to make understanding goals explicit and accessible as a focus for professional development.

The CCDT, with its interactive workspaces and attached message boards, is also a technology to support performances of understanding, ongoing assessment, and reflective collaboration among professional educators. The tool helps users draft curriculum designs that develop and demonstrate teaching for understanding with new technologies. With multiple examples of ways to apply these principles, the CCDT and other resources in the ENT Village help users reflect on their own work in relation to explicit criteria. They also support collaboration and dialogue with colleagues and coaches who participate in ongoing assessment of curriculum plans. Within this on-line digital environment, designers can easily revise curriculum plans, in light of ongoing assessment and feedback.

The ENT Village explicitly aims to engage its members in active, ongoing collaboration and reflection with like-minded colleagues. Those who enroll in the ENT Village receive an electronic "backpack" that organizes their work, such as designs they have created with the CCDT. The ENT Library contains links to a wide range of resources supporting the integration of new technologies, and users may place the resources that interest them in their ENT backpack. Members may also contribute to the library of resources, publish their curriculum designs, and participate in electronic forums in the ENT Meeting House. In these ways, members of the ENT Village help to create, as well as benefit from, a community of reflective colleagues.

WIDE World On-Line Professional Development

Thousands of users have registered in the ENT Village, but few became deeply engaged in using its resources. Busy teachers need more structure and support in order to collaborate on learning about teaching for understanding with new technologies. Desire to provide sustained and intensive support for these processes led a group of colleagues at the Harvard Graduate School of Education to develop an on-line professional development program. WIDE World (http://wideworld.pz.harvard.edu) builds on the resources of both the ENT Village and a similar Web site called ALPS (Active Learning Practices for School at http://learnweb.harvard.edu/alps). The ALPS site, originated under direction from David Perkins, helps educators apply a range of educational practices developed at Harvard's Project Zero, including Teaching for Understanding, applying Howard Gardner's theory of multiple intelligences, and Perkins's "Thinking Skills" curriculum.

WIDE World on-line courses take advantage of networked technologies to provide professional development that explicitly models the principles of Teaching for

Understanding. Instructors of WIDE courses design the syllabus with guidelines that help them to enact framework principles, for example, clear understanding goals, course activities that include performances of understanding, and ongoing assessment with coaching and feedback. WIDE's on-line course environment allows instructors to organize course materials in ways that explicitly highlight understanding goals. Learners in WIDE courses are clustered into study groups of ten to twelve individuals or small teams. Each study group works with a coach who helps the group develop comfort and trust, provides tailored encouragement and assessment for each member, and supports peer interaction. This social structure of WIDE courses supports the element of reflective, collaborative community and distinguishes this form of on-line professional development from self-guided tutorials.

Unlike more academic courses that might emphasize conceptual knowledge, WIDE courses focus on supporting learners in actively applying the ideas from the course to their own teaching. In other words, courses focus on performances of understanding, not simply providing training, distributing information, or building content knowledge. For example, in the WIDE course "Teaching to Standards with New Technologies," learners gradually refine or design a curriculum unit that integrates new technologies to support principles of Teaching for Understanding. Participants use the on-line CCDT, with its built-in supports and resources, as a structured environment in which to build their design. As participants draft each element of their curriculum design, coaches model ongoing assessment strategies and support participants in providing peer assessments. Through incrementally devising and revising a curriculum design that integrates new technology, participants experience and apply the process of Teaching for Understanding. Some

WIDE Courses

- WIDE courses focus on supporting learners in actively applying the ideas from the course to their own teaching. In other words, courses focus on performances of understanding, not simply providing training, distributing information, or building content knowledge.
- WIDE courses rely primarily on asynchronous communications, so participants can take time to think about their responses to assignments and prompts and choose their words carefully to express their meaning.
- Messages posted in electronic forums are organized into topics or threads and archived. This enables learners to read back over a series of comments, taking note of alternative perspectives, detailed explanations, and the evolution of ideas.

assignments in WIDE courses require participants to practice new strategies with their own students and to post reflections about how the process worked.

Explicit application of the principles of Teaching for Understanding is a hallmark of WIDE courses. Ongoing assessment with feedback from peers and coaches is carefully supported. In the "Teaching to Standards with New Technology" course, as participants read about effective ways of teaching for understanding with new technologies, the instructor gradually synthesizes these approaches into a rubric for assessing curriculum designs. Participants apply this rubric as they gradually design a curriculum unit around the elements of the framework. They work in a small group with a coach who uses the rubric to provide feedback on each element, as participants design their curriculum units. Seeing the coach model effective strategies for conducting constructive ongoing assessment helps participants begin to see how assessment can lead to improving performance.

Coaches also help participants learn how to conduct self-assessment and peer assessment. Teachers are often not comfortable offering feedback to one another. They may say things like, "I don't know anything about her situation, so how could I possibly offer her useful advice? It would be presumptuous to assume I have anything constructive to suggest." Nevertheless, teachers stand to benefit enormously from peer assessments and consultation. By using clear criteria to examine curriculum designs, they come to understand the criteria better. Studying the plans developed by fellow professionals often gives teachers good ideas that they can adapt for their own plans. And teachers often draw on their own expertise to make excellent suggestions to colleagues.

Recognizing both the aversion to peer feedback and the great advantages of it, WIDE courses provide specific support for this work. They may begin by helping participants view their own work in relation to explicit assessment criteria and then guiding them to ask for assistance around an area where they think their work could be improved. When teachers have directly asked for help, their colleagues are more willing to provide it. WIDE courses also adapt approaches developed by colleagues at Project Zero (http://www.pz.harvard.edu), an educational research and development center based at the Harvard Graduate School of Education, including strategies for looking at student work and conducting assessments that improve understanding. Culling from their work, some WIDE courses teach participants how to apply a "ladder of feedback" as they examine a colleague's work.

The "ladder of feedback" peer assessment process begins with questions of clarification so that the person conducting the assessment is sure he or she understands what the producer of the work intended. Then the peer assessor identifies strengths in the work, partly to bolster confidence but also to be sure that the good parts of the design are not lost as subsequent improvements are made. Finally, peer assessments include some suggestions (perhaps couched tentatively to acknowledge lack

TIP

Teachers stand to benefit enormously from peer assessments and consultation. By using clear criteria to examine curriculum designs, they come to understand the criteria better.

of complete information) about how the work might be improved. All of this analysis is framed in relation to the public assessment rubric or list of criteria that coaches have been using and that participants have applied in their own self-assessments.

Professional development aimed at helping teachers learn how to teach for understanding with new technologies should apply new technologies to help the participants gain that understanding. Several features of networked technologies encourage collaborative, reflective exchange. On-line communication is written rather than oral. As a result, participants can formulate their ideas and express them more deliberately than oral communication allows. WIDE courses rely primarily on asynchronous communications, so participants can take time to think about their responses to assignments and prompts and choose their words carefully to express their meaning. Messages posted in electronic forums are organized into topics (or threads) and archived. This enables learners to read back over a series of comments, taking note of alternative perspectives, detailed explanations, and the evolution of ideas. Reviewing the process of learning in this way can help teachers become more reflective about principles in relation to practice and their own process of development.[5]

Networked technologies also allow participants and coaches in a learning community to support one another from a distance. WIDE World recruits coaches from those who have successfully completed courses. WIDE offers an on-line course for prospective coaches to prepare them in providing social, pedagogical, and technological support for learners. Prospective coaches serve as apprentice coaches by taking an on-line course a second time while focusing on the work of an experienced coach. This method of recruiting and preparing coaches not only makes WIDE a scalable approach but also offers experienced teachers a professional ladder for developing new professional roles without removing them from their own teaching setting.[6]

Planning and Teaching Professional Development with New Technologies to Transform the Culture of Teaching

Providing the professional development and support that is necessary to improve teaching and learning with new technologies is neither quick nor simple. It goes far beyond merely giving teachers access to new technology or providing technical training. It encompasses much more than generating banks of "best practices" and making these resources available for teachers to search and adapt. Such examples may build teachers' awareness of the possibilities that new technologies offer. This is a useful first step, but most teachers also need guidance and support in learning how to integrate such practices into their own teaching. Interpersonal exchanges with a coach and peers are important supplements in helping teachers translate knowledge into practice. Furthermore, even the best-designed professional development activities will not accomplish their full potential unless teachers work in a context that helps them directly apply what they have learned. Educational systems that aim to improve teachers' and students' performances with new technologies must couple professional development with institutional and administrative support in the context of classroom work.

Meeting Teachers' Evolving Needs and Interests

Many studies of the process of integrating new technologies in schools show that teachers' needs and interests evolve as they work with new tools and teaching strategies; teachers require different forms of assistance over time.[7] Initially, developing teachers' awareness of "images of possibilities" may be accomplished by showing teachers examples of ways that colleagues are working with technologies. Giving teachers access to technologies that they can use for their own professional purposes and teaching them how to use these "productivity tools" can help teachers appreciate the potential of technology. Teachers may begin to develop technical expertise by using technologies to prepare curriculum materials, document student progress, and communicate with parents and colleagues.

Learning how to integrate technology into the classroom requires attention to multiple additional dimensions besides mere technical training. Effective integration of technologies in education requires teachers to modify curriculum and often entails significant changes in pedagogy, assessment, ways of orchestrating interactions in the classroom, and modes of collaboration both within and beyond the classroom. Professional development must guide, coach, and support teachers throughout their efforts to change these varied dimensions of their practice.

Teaching for understanding with new technologies involves multiple shifts of focus from typical classroom practices. Attention shifts from covering the curriculum to building students' capacity, from what teachers teach to what learners understand, and from what the teacher does to what students do. To foster these shifts, teachers must help their students take on new responsibilities. Students must become actively engaged in thinking and in working to apply their knowledge in creative ways. Learners must

Educational systems that aim to improve teachers' and students' performances with new technologies must couple professional development with institutional and administrative support in the context of classroom work.

become willing to share their draft work, to participate in thoughtful assessments of early drafts, and to contribute constructive suggestions about how to improve performances and products. To help students take on these new roles, teachers share their goals and assessment criteria explicitly and publicly. Teachers may even engage students in the process of defining goals and assessment criteria. Teaching for understanding depends on developing norms of respect, collegial exchange, dedication to thinking, and willingness to risk growing so that all the members of a learning community share responsibility for teaching and learning. In short, teaching for understanding depends on building a robust culture for learning.

Professional development aimed at promoting teaching for understanding must foster this same kind of culture of learning among teachers. Experienced professionals take pride in their expertise and may not feel comfortable calling their accustomed assumptions and practices into question. In schools where norms of privacy prevail, teachers usually avoid sharing "first draft" ideas and designs that are not fully polished. Building trust, celebrating those who risk change, and honoring feelings of uncertainty are all part of cultivating a culture of learning. On-line professional development must pay particular attention to developing a supportive "social presence" among learners who do not have access to the visual and verbal cues that can signal warmth and invite the development of trust.[8]

As teachers become more experienced in using new technologies to teach for understanding, their interests in curriculum work and commitment to collaboration evolve. They may want to engage in deeper analysis of understanding goals, refinement of curriculum designs that focus on higher-order thinking, and design of social relationships that promote peer assessment. As teachers gain familiarity with these strategies, they usually welcome opportunities to exchange ideas with others who are similarly immersed in these practices. Some teachers will also want to support colleagues who are at an earlier phase of development. Providing means for teachers to serve as part-time mentors or coaches for colleagues is a way to accomplish multiple purposes. It extends the roles of veteran teachers by providing new forms of professional experience without necessarily taking them away

Building a Learning Culture for Teachers

- Teaching for understanding depends on developing norms of respect, collegial exchange, dedication to thinking, and willingness to risk growing so that all the members of a learning community share responsibility for teaching and learning.
- Building trust, celebrating those who risk change, and honoring feelings of uncertainty are all part of cultivating a culture of learning.
- It requires an atmosphere that encourages learners to put forth draft ideas, to give and receive constructive suggestions about improving work, and to embrace learning as an ongoing process of inquiry, not just as a process of raising test scores.
- Teaching for understanding depends on a deep commitment to intellectual effort and to high standards for the work that teachers and students produce.
- Administrative support for teachers who take initiative to deeply examine, understand, and redesign their practice is invaluable.

from teaching. And it cultivates instructional leaders who enrich the network of support for fellow teachers who are experimenting with new approaches.

Creating Supportive Contexts

Direct support for teachers is necessary but not sufficient to achieve significant educational leverage from the introduction of new technologies. The context in which teachers work, including its technology, organizational structure, politics, collegial atmosphere, and educational culture, significantly affects what teachers are able and motivated to do with their students. Unless these contextual dimensions are at least somewhat supportive of teaching for understanding with new technologies, teachers are unlikely to sustain such practices.[9]

The amount of technology available to teachers obviously shapes what teachers can do in their classrooms, but the simple ratio of computers-to-students is not the most important factor. The type and distribution of technology also matters, because different configurations may be suitable, depending on what teachers want to do. A projector that displays the work of one computer for the whole class may cost less than adding more computers and may be much more useful for educational purposes. Inexpensive, handheld devices or portable computers that can be packed up and moved from one class to another may be more useful than a lab full of stationary desk-top computers. Connectivity to the Internet is also an influential factor shaping the educational potential of computers. Wireless access may be easier to install than hard-wired infrastructure. Technical staff people who maintain computers and provide technical assistance to teachers and students may be a more important investment than additional purchases of technology.

Organizational structures also shape what teachers and students are able and willing to do in their classes.[10] Time is the engine of most schools. Class schedules, as well as their intersection with other schedules such as access to computer labs, affect the kinds of activities that teachers can plan. Projects that involve significant time for inquiry (gathering and analyzing data) may require more time than the usual forty-five-minute class period. If students have to chop their work into short segments, they may spend more time assembling their materials than actually thinking about what they have found. Teachers also need time to plan, set up, and coordinate lessons that go beyond the usual lecture-and-seatwork format. If they wish to collaborate with other teachers, they need simultaneous planning periods with those colleagues and coordinated class schedules. Some flexibility around time can be arranged if administrators will make teaching for understanding with new technologies a priority, enabling teachers to devote their efforts to this goal instead of other activities like committee meetings, required professional development workshops on another topic, or administrative duties.

Of course, political forces are a major influence in all organizations. Principals set the tone for the allocation of power in school buildings, just as superintendents and the school board do in school districts and union leaders do among the faculty. If these political leaders are aligned in support of shared goals, the stage is set to advance their priorities. If, as is often the case in school systems, key leaders are pulling in different directions, teachers feel buffeted by crosswinds. Clarity and alignment around the educational agenda is most important in determining whether teachers can work in a sustained and focused way on an approach like teaching for understanding. If priorities are splintered or short-lived, such initiatives cannot garner the multilayered, consistent support that they require.

Teaching for understanding is an educational approach based on openness about goals and collegial reciprocity among all the members of a community of learners. It works best when all participants share responsibility for teaching and learning. It requires an atmosphere of trust that encourages learners to put forth draft ideas, to give and receive constructive suggestions about improving work, and to embrace learning as an ongoing process of inquiry, not just a process of raising test scores. If these values pervade the culture of a school, they are apparent in staff meetings, in the way children are treated in the classroom and on the playground, and in the relationship between teachers and parents. In such a context, the norms underlying teaching and learning for understanding are consistent from one class to the next and from one year to the next. Children learn how to share responsibility in a community of learners over time. If these values do not prevail, teachers who wish to cultivate them in their own classrooms have to devote enormous effort and care in mentoring their students around issues of risk taking, intellectual effort, and collegial reciprocity.

Along with these norms of collaboration, teaching for understanding depends on a deep commitment to intellectual effort and to high standards for the work that teachers and students produce. The student work displayed on the walls indicates the values of a school. Uniform cutouts, with small variations in the markings students have made, signal that neatness trumps invention. Perfectly formatted final essays that may be only a paragraph long, lack content, and closely replicate a template suggest that teachers regard word processors as devices for making work look professional, not as tools for developing writing that is vivid, compelling, detailed, and unique. In contrast, portfolios of student work showing the evolution of a project from early drafts through multiple revisions to a final polished form demonstrate a school's priority on learning as a process. Presentations in which students share their understandings and explain how they used varied tools to develop their learning demonstrate a school's commitment to teaching for understanding with new technologies.

Teachers and their students alone cannot generate the contextual support that is needed for such accomplishments. They require support from administrators and leaders of other constituencies that influence schools: parents, faculty, staff, businesses and other taxpayers, and policymakers at all levels from local to national. Administrative support for teachers who take the initiative to deeply examine, understand, and redesign their practice is invaluable. Professional development that helps teachers build knowledge, skills, beliefs, and relationships is essential in supporting teaching for understanding with new technologies. But these activities must be complemented with commitment from all the other stakeholders in schools if new technologies are to make a significant contribution in preparing students to be creative and productive citizens in their local and global communities.

Questions for Reflection

1. Whatever your role may be in an educational system, how can you help to promote professional development activities and contexts that support teaching for understanding with new technologies?

2. How might you take advantage of new technologies to develop and sustain such activities and supportive conditions?

NOTES

1. Wiske, M. S., and Perkins, D. "Dewey Goes Digital: Scaling Constructivist Pedagogies and the Promise of New Technologies." In C. Dede, J. P. Honan, and L. Peters, (eds.), *Scaling Up Success: Lessons Learned from Technology-Based Educational Innovation.* San Francisco: Wiley, in press.

2. These recommendations about effective professional development are phrased to highlight their correlation with the elements of teaching for understanding, but the recommendations are entirely consistent with research and guidelines on successful professional development practices. See, for example, the following titles on the Web: *Bridging the Gap Between Standards and Achievement: The Imperative for Professional Development in Education* (Elmore, R. F. Albert Shanker Institute, 2002; retrieved Nov. 4, 2003, from http://www.ashankerinst.org/Downloads/Bridging_Gap.pdf); *No Dream Denied: A Pledge to America's Children* (National Commission on Teaching and America's Future, Jan. 2003; retrieved July 30, 2003, from http://www.nctaf.org/dream/report.pdf), and *E-Learning for Educators: Implementing the Standards for Staff Development* (National Staff Development Council/National Institute for Community Innovations, 2001; retrieved Nov. 24, 2003, from http://www.nsdc.org/library/authors/e-learning.pdf).

3. Little, J. W., and McLaughlin, M. W. (eds.). *Teachers' Work: Individuals, Colleagues, and Contexts.* New York: Teachers College Press, 1993.

4. For a rich discussion of the nature and value of professional discourse communities, see "What Do New Views of Knowledge and Thinking Have to Say About Research on Teacher Learning?" (Putman, R., and Borko, H. *Educational Researcher,* 2000, *29*(1), 4–15).

5. For thoughtful analyses of the value of on-line communication in fostering reflective thinking, see *E-Learning in the 21st Century: A Framework for Research and Practice* (Garrison, D. R., and Anderson, T. London: RoutledgeFarmer, 2003).

6. For more on the coaching process as part of a sustainable professional development process, see "New Technologies to Support Teaching for Understanding" (Wiske, M. S., Sick, M., and Wirsig, S. *International Journal for Educational Research,* 2002, *35*(5), 483–501).

7. Sandholtz, J., Ringstaff, C., and Dwyer, D. *Teaching with Technology: Creating Student-Centered Classrooms.* New York: Teachers College Press, 1997. See also, *The Connected School: Technology and Learning in High School* (Means, B., Penuel, W. R., and Padilla, C. San Francisco: Jossey-Bass, 2001).

8. Garrison, D. R., and Anderson, T. *E-Learning in the 21st Century: A Framework for Research and Practice.* London: RoutledgeFarmer, 2003. The authors devote Chapter 5 to a discussion of "social presence." The importance of developing communities of learning in on-line courses is a focus of *Distributed Learning: Social and Cultural Approaches to Practice* (Lea, M. R., and Nicoll, K. London: RoutledgeFarmer and the Open University, 2002).

9. Elmore, R. F. *Bridging the Gap Between Standards and Achievement: The Imperative for Professional Development in Education.* Washington, D.C.: Albert Shanker Institute, 2002; Little, J. W., and McLaughlin, M. W. (eds.). *Teachers' Work: Individuals, Colleagues, and Contexts.* New York: Teachers College Press, 1993.

10. Aspects of school culture, structure, and politics that impede effective uses of technology are main themes in *Oversold and Underused: Computers in the Classroom* (Cuban, L. Cambridge, Mass.: Harvard University Press, 2001).

Learning for the Future

The central arguments of this book are that (1) learners need to develop the capability to perform with their knowledge, (2) learning is an active and social process of inquiry and interpretation, and (3) teaching is a process of cultivating learners' capacities to develop and apply their understanding creatively in the world. New technologies make their best contribution to education when they support these kinds of teaching and learning. Teaching for understanding with new technologies is a complex process that depends on coordinated support and sustained professional inquiry. The goals of this book will be realized if readers use these ideas to spur continual inquiry and collaboration toward ongoing improvement of both the principles and the practices of teaching for understanding.

This kind of collaboration within professional communities of thoughtful educators is especially necessary, yet particularly difficult, within the current contexts of global politics and educational policies in the United States. A short review of these circumstances sets the stage for an urgent call to action.

Schools for Today and the Future

The public education system is the generator of a democratic society. If a government of the people and by the people is to be for the people, those people must be informed and responsible enough to vote intelligently. They must elect representatives and make choices on referenda that guide society toward the good. Citizens of a democracy invest in public schools partly to prepare the next generation to be effective stewards of a democratic way of life. Whatever else the schools may be expected to accomplish, they must also fuel the civic engine of democracy.

During the past century, commitment to universal public education has brought an increasingly diverse range of students into the schools. These varied learners create both opportunities and responsibilities. There is opportunity for teachers and students to learn from and develop respect for one another's diversity, whether that stems from their varied linguistic, cultural, and ethnic backgrounds or differences in their abilities and interests as individual learners. Diverse students also generate new responsibilities for teachers to support learners with a wide range of needs, including many who are directly and indirectly affected by stressful conditions in families, neighborhoods, and the world. Above all, schools and teachers bear responsibility to provide high-quality education for *all* students.

Defining *high-quality education,* always a topic worthy of debate, becomes particularly problematic in an era of rapid change. Developments in information, communication, and transportation technologies have dramatically altered the effects of distance in time and space. These changes raise questions about what and how and where people need to learn. More than at any previous time in human history, the notion of a global village is not an abstraction but a reality. Whether or not people understand and appreciate the extent to which their lives are entwined with others around the world, they are, in fact, enmeshed in a complex web of interacting global systems. Local decisions, for example, about how to use energy, water, food, and other resources, how to resolve conflicts, and how to organize collaborative action generate global consequences.

Conceptions of responsible citizenship expand as the future of planet Earth increasingly requires global cooperation in order to preserve the health of the environment and the safety of its species. In an interdependent global society, understanding one's own culture and contributing to one's local or national community is essential but not sufficient. Today's children must also learn why and how to understand and cooperate effectively with people who may be quite different from themselves in countries and cultures around the world if they are to preserve and extend democracy and ensure the welfare of the planet.

Teaching for Today and the Future

Traditional images of teachers depict a self-contained figure of authority transmitting facts and formulas or directing students in recitation of the material previously taught. In this view, teachers transfer the canon of accepted knowledge by covering the curriculum. Accordingly, teachers may be held accountable by tests that measure whether their students have learned the required facts and formulas. This view of instruction is currently being reinforced by a movement focused on standards-based curriculum and required achievement tests that is sweeping through the educational system in the United States and some other countries. Advocates of these policies believe they hold schools and teachers accountable for providing high-quality education to all students. Other consequences of required curriculum and tests may not be so beneficial, however; they might drive teachers to cover curriculum at an overly rapid pace and to drill their students on isolated test items rather than build students' understanding of what they learn.

The instruction-as-delivery metaphor has never been sufficient to encompass all the dimensions of a teacher's role, however, and is obviously inadequate for preparing students to succeed in the twenty-first century. Teachers have always had to learn about their pupils in order to successfully foster students' learning. They have always had to understand their subject matter well enough to explain it in several ways, not just to deliver a defined curriculum.

The shortcomings of delivery-style instruction are more apparent than ever in today's educational context, for several reasons. First, the nature and amount of information and the interconnectivity of the world is changing and growing dramatically with the advent of new technologies. Certainly, students need to learn about some of the knowledge that their predecessors have invented and discovered. But they must also learn how to be astute navigators in a sea of ever-expanding information. They must learn how to sift the relevant from the irrelevant, as they assess bias and connect separate-but-related bits of information. Instead of learning lots of isolated facts and formulas, students need to understand relationships among key concepts, modes of disciplined reasoning, and thoughtful habits of mind so that they can recognize patterns, make sense, reason analogically, derive generalizations, and apply lessons learned in one setting to other contexts. Indeed, students must also learn how to contribute and co-construct knowledge, and these capabilities must be directly cultivated from the beginning of their lives. Teachers aiming for such goals cannot simply transmit knowledge generated by others; teachers must also prepare students to think for themselves. In order to do that, teachers must also be required and encouraged to think for themselves, not merely transmit the required curriculum.

Second, in a diverse and rapidly shrinking and interdependent world, future citizens of a global democracy must learn how to understand and cooperate with one another. A central component of their education must be not just building up the store of their individual knowledge but developing the capacity to interact and collaborate effectively with other people. Today's students need to understand how and why people in other parts of their town, country, and planet are different from themselves and what they all have in common. They need to appreciate diverse perspectives, not just tolerate them. And they must be skillful in forging collaborative relationships to formulate and address projects that are too complex for any individual or small group to accomplish. In order to develop their students' capacities in these ways, teachers must embody and model the same kinds of perspectives and abilities. Teachers must be able to cultivate collaborative communities within their classrooms and to help their students form relationships that go beyond their classroom walls.

A third reason that the delivery conception of teaching is inadequate derives from research on effective learning. During the past thirty years, a growing body of research on thinking and cognition has demonstrated that learners must be actively engaged in changing their minds, not just assisted in storing new information on the empty shelves in their minds. Students come to school with preconceptions about science, history, mathematics, and other subjects. These ideas may be resilient misconceptions that are difficult to change. Simply telling students a different idea does not usually alter their thinking. Teachers need to help students actively construct new understandings through acknowledging initial conceptions, making predictions, analyzing evidence, and thinking for themselves about explanations and mechanisms that could account for the evidence. Of course, students can learn some information, knowledge, and skills from reading or being taught. But in order to develop broad, flexible, durable, and transferable "usable" knowledge, students must play an active part in developing their own understanding.

Helping students construct their own understanding requires teachers to do much more than deliver instruction.

Helping students construct their own understanding requires teachers to do much more than deliver instruction. Teachers must understand their students and take account of students' particular interests, knowledge, and misconceptions as teachers develop or revise curriculum plans. Teachers must possess a deep and flexible understanding of their subject matter in order to help diverse students make sense of their lessons and build their own capacity to think. Teachers must design and mediate activities that engage students in actively developing their understanding through inquiring, reasoning, and applying what they know to problems and projects. Indeed, if learning is viewed as a process of constructing understanding instead of absorbing information, some fundamental issues of intellectual responsibility and authority must be negotiated. Learners need to take on responsibility for creating knowledge and sharing their ideas. Teachers must themselves be active learners, who model the struggles and satisfactions of not yet knowing and who are willing to learn from others, including their students. Both teachers and pupils share responsibility for teaching and learning when understanding is the goal.

Teachers who are preparing their students for today's world and the future must be educational professionals, not mere channels for delivering instruction. They must be able to exercise judgment about what to teach, how to teach, and how to assess their students' progress. Of course, as policymakers and school administrators set guidelines and standards, teachers must make efforts to fulfill these priorities, but that is not accomplished solely by covering only required curriculum or by drilling students on items from a standardized test. In order to fulfill their responsibilities to their students, teachers must exercise their own judgment on curricular design and implementation, be continuing learners about their subject matter and their craft, and build collaborative relationships that connect their classroom with the world beyond.

To fulfill their responsibilities to their students, teachers must exercise their own judgment on curricular design and implementation, be continuing learners about their subject matter and their craft, and build collaborative relationships that connect their classroom with the world beyond.

The Role of New Technologies

Just as the development of new technologies is transforming the world, it influences school curricula, practices, and tools. Since the early 1980s, many school systems have invested heavily in purchasing computers and networking technologies to use for both administrative and educational purposes. This trend began in the developed world and is spreading to developing nations as the cost of technologies declines, and new inventions such as wireless networks enable countries to "leapfrog" over limitations in hardware and infrastructure. In many places, a general sense that modern technologies will make schools competitive and that students must get "on computers" in order to be prepared for the future prompt decisions to purchase hardware without any clear educational planning. In these cases, the computers often remain in their unopened boxes or sit idle because school people have not devised ways of putting the new technology to good use.

When schools or educational planners do get around to asking how new technologies might relate to their educational goals, several categories of proposals usually arise. The proposals tend to advocate that computers be used for one or more of the following purposes: tutor, tool, or tutee. The notion of *computer as tutor* builds on the basic idea of teaching machines programmed to promote learning. Examples include drill-and-practice activities structured like electronic workbooks and tutorials that may include animated simulations and alternative pathways for learners to explore. Computers as *tools* can extend and enrich the repertoire of materials and artifacts that teachers and students use to promote learning. Equipped with appropriate software, computers can serve as successors to typewriters, books, encyclopedias, adding machines and other kinds of calculators, thermometers and other scientific instruments for gathering and analyzing data, card catalogues, telephones, and televisions. Recent technologies such as handheld computers use wireless transmitters to connect with one another and may be augmented with cameras, recorders, and data probes. Some people believe these tools not only can augment traditional classrooms but can radically transform the physical and virtual spaces where learning takes place. Treating computers as *tutees* puts teachers and students in the position of "teaching" the computer by programming it to serve the author's purposes. Teachers and students who use technology as a tutee become producers of new technologies to serve their purposes, not just consumers or passive recipients.

Just as there are varied conceptions of the role that computers might play in the educational process, there is also a range of ideas about the relationship between technology-enhanced activities and the usual curriculum. Technology may be used as an *aid* to improve the efficiency or effectiveness of the existing curriculum, as when graphing calculators are integrated into algebra classes. It may become an *addition* to the curriculum, as happened in many schools where computer classes and teachers were added to the educational program, sometimes substituting for other classes, such as the arts or physical education. In more wide-reaching and fundamental ways, new technologies may be a *transformer* of the traditional curriculum, enabling teachers and students to learn about topics or learn in ways that were not feasible or possible without the new technology. Using the Internet to teach with accurate, continually updated data sets, to engage students in collaborative creation of hypermedia products, and to connect students with experts and peers outside their classrooms in collaborative construction of knowledge are examples of transformational applications of educational technologies.

Within this welter of possibilities are many ways of using multimedia, interactive, hyperlinked, and networked technologies to enhance and enable the kinds of educational experiences we described in the previous section. As educators become more familiar with the potential of new technologies, they become more able to integrate them fluently into classroom practice and school processes so that distinctions blur among tool, tutor, or tutee and among aid, addition, or transformer. In classrooms like Kristi Rennebohm Franz's, new technologies are pervasive resources seamlessly and continually woven into the activities of teacher and students.

Yet such classrooms are still uncommon. Many would argue that computers in schools have fallen far short of their promise. In many schools, universities, and other educational settings, computers are underused, minimally used, or misused rather than being employed in ways that make a significant contribution to teaching and learning.[1]

The shortfall between the hopes and the reality of computers in most educational settings is primarily caused by a lack of clear, focused plans for using the technology in ways that will significantly improve or transform education. This book offers a means of bridging this gap by presenting a systematic framework and process for making reasoned decisions about whether, when, and how to take advantage of new technologies to improve teaching and learning for understanding. Ideally, the process of preparing teachers to use new technologies in these ways is itself an experience of learning for understanding through ongoing collaborative inquiry by educators.

Aligning Components of Teaching for Understanding

The Teaching for Understanding framework presented in this book is best regarded as a set of guidelines for continually refining good teaching. Reading separate chapters about the different elements of teaching for understanding may obscure the importance of connecting these elements coherently and cyclically. Understanding goals become the basis of defining criteria for ongoing assessment of performances. As teachers devise rubrics for assessing their learners' performances, they review their goals to make sure that all the key goals are reflected in the rubric. Indeed, some teachers discover that the qualities they hope to see in their students' final products reflect some goals that were not included in their stated understanding goals. Such a discovery may prompt a revision in the goals and a redefinition of the generative topic, as well as a change in the assessment rubric.

Understanding goals must also correlate with performances. Each major learning activity and performance of understanding should help students develop and demonstrate understanding of one or more target goals. Explicitly linking each major student performance with understanding goals and subgoals is a good way to check that the work students are asked to do directly fosters their learning about important dimensions of understanding. If the connections cannot be made, the assignment may not be worthwhile or the goals may need to be revised.

Performances of understanding should also be aligned with the generative topic and with ongoing assessment. The guidelines for generative topics remind teachers to select materials and to design performances that allow learners to connect their study to their own experience and interests. A rich array of performances allows students to approach a topic through several entry points and to learn and express learning with multiple intelligences.

All major performances should be assessed in ways that generate useful information about ways to improve. Assessments may be accomplished informally in ways that do not require a disruption in learning. Teachers may conduct assessments by listening to students as they participate in discussions or make presentations to the class or by observing students as they collaborate in small groups. Slightly more formal assessments may be conducted if teachers complete a short checklist as they observe students. Learners may be asked to select work that exemplifies the evolution of their understanding to include in a process portfolio, using a rubric to guide their selections. Embedding assessment within the educational experience requires little time away from the ongoing process of learning and, in fact, contributes to helping students develop understanding. As a general guide, assessments become more formal, explicit, and time consuming as students progress toward their culminating performance. During introductory "messing about" performances, teachers may observe students to gauge their interests and beginning levels of understanding. During "guided inquiry" performances, rubrics may still be evolving, and the assessment process focuses on generating suggestions for improvement. Culminating performances should result from multiple rounds of assessment, coaching, and feedback, so that the final assessment demonstrates students' best understanding and yields few surprises for either students or teachers.

Cultivating a culture of reflective, collaborative inquiry is an ongoing, integrated process that supports all elements of teaching for understanding. Developing students' capacities to engage in deliberate reflection about their own work is promoted by making understanding goals explicit and is necessary if students are to take part in useful self- and peer assessments. Building students' skills for collaboration helps to make learning more generative and enriches the range of performances through which students can develop and demonstrate their understanding.

All in all, teaching for understanding is not a matter of attending to each element of the framework as a discrete component. Instead, it is an ongoing process of progressively refining and aligning these elements so that they interact coherently to promote focused understanding. The process is never complete. Teaching can always be improved, so long as teachers are willing to be learners.

Integrating New Technologies to Improve Understanding

The systemic integration of new technologies to improve understanding is also an ongoing process of inquiry. The extended case studies in previous chapters of this book emphasize the incremental process by which teachers learn about new technologies, develop expertise, and refine ways of using these tools to improve teaching and learning with their students. This gradual process also holds true for all the participants in an educational system, including administrators and policymakers, as well as teachers. In all organizations, people tend to use new technologies initially to replicate their accustomed practices, just as early movies replicated theater performances on a stage, before they gradually discover the new tools' more innovative possibilities.

People tend to use new technologies to replicate their accustomed practices, just as early movies replicated theater performances on a stage, before they gradually discover the new tools' more innovative possibilities.

Teachers' strategies for making use of a particular tool usually evolve as they try to integrate it into their practice. As learners become more adept with new technologies, they may demonstrate new approaches and accomplishments that teachers had not anticipated. As teachers discover new possibilities that technologies afford, they may modify their curriculum plans, shift their teaching strategies, and devise new roles and structures for members of their class. Students themselves may become a source of assistance in the classroom, as they learn how to solve their own problems and help their peers. Technologies themselves are continually evolving. Teachers who want to take advantage of these developments must find ways to keep learning about these emerging possibilities.

All these forces for evolution operating at the classroom level have counterparts operating at the level of a school building, district, or larger organizational system. Decision makers at each level of an educational system must evolve their understanding of the potential of new technologies, their skills in applying these tools, and their ways of organizing resources and people to take advantage of the potential. At the school-building level, choices must be made about what hardware and software to purchase, where to locate these resources, and how to develop roles to provide technical assistance, maintenance, and support for educational integration. Similar kinds of choices must be made for school districts, state departments of education, and federal or national education agencies. As with teachers in the classroom, choices about technology at each level of the educational system will lead to better educational results if they are guided by a clear educational agenda.

Yet the educational agenda itself must also be regularly refined as the technologies change and the understanding about their educational potential develops. Decisions about how technologies can best be integrated into educational settings to improve learning require continual redefinition of goals, analysis of results, and reflection toward next steps. In order for teachers to participate in ongoing reflection about integrating new technology, they need to be supported by people in layers of surrounding contexts who are themselves engaged in similar inquiry. Ideally, the policies and practices in these different layers of the educational system are coordinated so that their combined effects encourage coherent, sustained inquiry aimed at fostering understanding.

Nurturing a Culture of Learning

None of the recommendations in this book can be met simply by following directions. The Teaching for Understanding framework supports a process of inquiry and invention to be undertaken by thoughtful and courageous educators. It is most valuable when treated as an invitation to reflect, define goals, design new practices, assess how they work, and then reconsider both the practice and the guiding principles. Those who have worked with the Teaching for Understanding framework have adapted or translated its terms to fit the priorities and ongoing initiatives in their own situation. One educator created a kind of glossary or modern Rosetta stone, correlating terms from his country's reform agenda with the elements of Teaching for Understanding and their particular criteria. This kind of translation of local priorities into the terms of the framework is an inventive approach to using this model as a spur for learning, not as a rigid recipe or rubric.

In a world that is changing rapidly, people must be continuing learners. Public education systems must engage administrators and teachers, and as well as students, in a process of learning how to learn. This process depends on participation in reflective, collaborative communities that debate educational goals and develop coherent strategies for achieving them. The Teaching for Understanding framework and new technologies provide a language and tools to promote such debates and to galvanize coherent collaboration. They offer a structure and enriched resources for people in schools, educational systems, and local and global communities to create schools that truly prepare children to reach their fullest potential in today's complex world.

NOTE

1. Cuban, L. *Oversold and Underused: Computers in the Classroom.* Cambridge, Mass.: Harvard University Press, 2001.

Glossary

Generative topics (one of the five elements of the Teaching for Understanding framework): Are curriculum topics that are connected to multiple important ideas within and across subject matters authentic, accessible, and interesting to students; fascinating and compelling for the teacher; approachable through a variety of entry points and a range of available curriculum materials and technologies. Generative topics have a "bottomless" quality that generates and rewards continuing inquiry. This element of the Teaching for Understanding framework helps educators address the question, What topics are worth teaching for understanding?

New technologies: May include any new tools for information and communication beyond the ones traditionally used for teaching and learning. Examples include video recorders and players, graphing calculators, computers equipped with any kind of software, digital probes linked to a display device such as a calculator or computer, the Internet with its World Wide Web of hyperlinked, multimedia Web sites, e-mail, and videoconferencing—resources that can be used to help students wonder, think, analyze, try to explain, and present their understanding. The main criteria are that the technology has significant potential to enhance students' understanding and is not yet part of the teacher's repertoire of educational tools.

Ongoing assessment (one of the five elements of the Teaching for Understanding framework): Is conducted frequently and generates suggestions for improving performances; is based on explicit, public criteria related directly to understanding goals;

may include informal assessments embedded within learning activities, as well as more formal and separate structures and products. Learners may conduct ongoing assessments of their own and their peers' work, in addition to assessments carried out by teachers, coaches, and others who provide feedback.

Performances of understanding (one of the five elements of the Teaching for Understanding framework): Develop and demonstrate students' understanding of target goals; require active learning and creative thinking to stretch learners' minds; build understanding through a sequence of activities from introductory "messing about" performances through guided inquiry to culminating performances, and engage learners with a rich variety of entry points and multiple intelligences.

Reflective, collaborative communities (one of the five elements of the Teaching for Understanding framework): Support dialogue and reflection based on shared goals and a common language; take into account the diverse perspectives of learners; promote respect, reciprocity, and collaboration among members of the community in communal, as well as individual performances.

Rubric: Lists the features of high-quality work and describes benchmarks for rating the extent to which a product meets these criteria. Criteria should correlate fully with understanding goals. Teachers who use the Teaching for Understanding framework often engage learners in developing such rubrics so that students understand the rubric and can use it to assess their own and their peers' work.

Targets of difficulty: Are typically difficult to teach and learn, are centrally important in the subject matter and the curriculum, and might be made easier to understand with new educational technologies. This term was coined by the Educational Technology Center at the Harvard Graduate School of Education to designate topics worth focusing on when designing ways of integrating new technologies with curriculum and instruction.

Teaching for understanding: A phrase often used to refer to a wide variety of educational approaches. Usually, it connotes some focus on meaningful learning and some strategies that involve students in active hands-on tasks with the goal of helping them make sense of their studies.

Teaching for Understanding framework (sometimes shortened to TfU): Refers to a particular framework or model based on collaborative research conducted at the Harvard Graduate School of Education. This framework includes five elements: (1) generative topics, (2) understanding goals, (3) performances of understanding, (4) ongoing assessment, (5) and collaborative, reflective communities. Together these elements outline a process and a set of criteria that guide teachers as they plan, implement, and assess strategies that support their students' understanding.

Throughlines: Used by some practitioners of the Teaching for Understanding framework to refer to long-term, overarching understanding goals, because they provide a focus for the entire arc of a course or a project, just as a throughline shapes an actor's performance in method acting.

Understanding: Is conceived in the Teaching for Understanding project as a capability to think and act flexibly with what one knows. This definition emphasizes understanding as a performance capability that goes beyond remembering what one has been taught or rehearsing routine skills.

Understanding goals (one of the five elements of the Teaching for Understanding framework): Clearly and publicly define what learners will come to understand; address multiple dimensions, including important knowledge, methods of inquiry and reasoning, purposes for learning, and forms of expression; are connected coherently so that lesson-level goals relate to long-term goals and to overarching goals, or throughlines.

Index